Olajide Jonathan Adamolekun

Reservoir quality controls and reservoir modelling of continental to transitional shallow marine sedimentary basins of Southern Nigeria

Reservoir quality controls and reservoir modelling of continental to transitional shallow marine sedimentary basins of Southern Nigeria

by
Olajide Jonathan Adamolekun

KIT Scientific Publishing

Karlsruher Institut für Technologie
Institut für Angewandte Geowissenschaften

Reservoir quality controls and reservoir modelling of continental to transitional shallow marine sedimentary basins of Southern Nigeria

Zur Erlangung des akademischen Grades eines Doktors der Naturwissenschaften (Dr. rer. nat.) von der KIT-Fakultät für Bauingenieur-, Geo- und Umweltwissenschaften des Karlsruher Instituts für Technologie (KIT) genehmigte Dissertation

von Olajide Jonathan Adamolekun

Tag der mündlichen Prüfung: 22. Dezember 2022
Erster Gutachter: Prof. Dr. Christoph Hilgers
Zweiter Gutachter: Prof. Dr. Michael Peter Suess
Dritte Gutachterin: Prof. Dr. Nevena Tomašević

Impressum

KIT Scientific Publishing

Karlsruher Institut für Technologie (KIT)
KIT Scientific Publishing
Straße am Forum 2
D-76131 Karlsruhe

KIT Scientific Publishing is a registered trademark
of Karlsruhe Institute of Technology.
Reprint using the book cover is not allowed.

www.ksp.kit.edu

Print on Demand 2023 – Gedruckt auf FSC-zertifiziertem Papier

ISBN 978-3-7315-1287-5
DOI 10.5445/KSP/1000156506

Acknowledgements

I would like to express my sincere gratitude to my advisors Christoph Hilgers and Michael Peter Suess for providing me with the necessary guidance and support at every stage of this work. Frequent discussions with Christoph Hilgers have been a great help in the proper formulation of ideas that greatly benefited this work. Beyond this research, most of my life has benefited from the training and support of Christoph Hilgers. I cannot thank you enough for making my research experience in Karlsruhe worthwhile. Michael Peter Suess particularly introduced me to a new workflow in reservoir modelling and uncertainty quantification, and supported me in overcoming challenges associated with a limited data set. Nevena Tomašević, is thankfully acknowledged for being the third referee of my doctoral thesis. Her thorough review and comments improved many parts of this work.

Words are not enough within the context of this thesis to thank Benjamin Busch who introduced me to diagenetic research principles and provided support and guidance throughout the execution process of this work. His internal review of my manuscripts has led to improvements in their contents and quality. Also, I am grateful for numerous hours of discussions on scientific writing and publishing ethics. Thank you for not getting tired of my many administrative questions also.

The support of Nicolaas Molenaar and Oladotun A. Oluwajana during the execution of this research are duly acknowledged. The discussions with Agnes Kontny are always inspiring and encouraging. This is duly acknowledged and appreciated. I acknowledged the assistance of Matthias Schwotzer and Felix Krause of the Mineral Interfaces research group, Institute of Functional Interface (IFG), Karlsruhe Institute of Technology (KIT), Campus North on X-ray diffraction measurements. Gabriel T. Olawuyi and James Bamigbose are thankfully acknowledged for their support during the fieldwork and core sampling.

Many thanks to the past Structural Geology and Tectonics (SGT) group team members at KIT for the pleasant and memorable experience I had. Alexander C. Monsees is particularly acknowledged for his support and assistance in many forms. He introduced me to the usage of the drilling equipment and also supported Benjamin Busch on my introduction to diagenesis and helped with administrative issues. Alex has been highly supportive of me right from my first day at Karlsruhe up till now (after he left KIT). Ivy Becker induced me to the usage of the AccuPyc II 1340 by micromeritics® and the air permeameter. Christina

Schmidt is acknowledged also for her support. She is always open to discussing and supporting to the best of her ability. Discussions with my present colleagues; Katharina Steiger, Jasemin Ölmez, Dennis Quandt, Felix Allgaier, Chaojie Cheng, Jonas Greve and Bruno Mendes are highly acknowledged.

Stephan Unrein and Martin von Dollen are greatly acknowledged for the thin-section preparation. I appreciate the basic skills and knowledge in thin-section preparation gained from Stephan Unrein. With Ulrike Brecht on board in SGT, administrative issues are perfectly handled with proper guidance. Her help, understanding and support are highly appreciated.

The German Academic Exchange Service (DAAD) and the Petroleum Technology Development Fund (PTDF) are greatly acknowledged for granting me a PhD scholarship for this research work. The Nigerian Geological Survey Agency (NGSA) and the Nigerian Bitumen Corporation are acknowledged for the provision of core samples used in this work. Shell Petroleum Development Company (SPDC) is acknowledged for the provision of seismic and well log data used for the modelling.

To my former university lecturers Matthew E. Nton, Michael A. Oladunjoye, Olugbega A. Ehinola and Olubunmi, C. Adeigbe who are always available for my many questions, especially on the highly debated stratigraphy of the Eastern Dahomey Basin, I say thank you all.

I thankfully acknowledge my mother, siblings, friends and other close relatives for their support and encouragement. Thanks for your invaluable care and love. To my loving wife; Oluwaseun O. Adamolekun, who became part of my journey towards the end of this work, I cannot appreciate you enough. Thanks for your support and understanding, and for quickly adapting to the life of a researcher. She is always interested in my daily achievements and my overall progress.

Finally, I owe this thesis to the God Almighty who by his love, favor, and mercy made this an accomplishment, despite numerous challenges.

Abstract

Continental to shallow marine depositional systems contain significant reservoirs worldwide. However, facies heterogeneity poses a difficulty in accurate prospect delineation, field development, and production strategies. Understanding of reservoir properties and their distribution and controls promise better exploration success.

The Eastern Dahomey and Niger Delta Basin in Nigeria are formed during the separation of the African and South American Plates in the Late Mesozoic, following the break-up of the supercontinent Pangea. Local and regional heterogeneity in lithofacies has made accurate location and prediction of reservoir rocks difficult in the Eastern Dahomey Basin in the west. Contrarily, the Niger Delta in the east is a highly petroliferous basin. The understanding of reservoir heterogeneity, reservoir quality controls, and improved 3-D models in this region can enable greater exploration success. Therefore, this study investigates the controlling factors of a reservoir within the framework of a 3-D static modeling of an onshore field in the Niger Delta Basin using seismic data and geophysical logs. It is shown how important the deposit quality is for the productive reservoir volume. In the following, this study focuses on the diagenetic history and deposit quality assessment of deposit rocks in the Eastern Dahomey Basin using petrographic, mineralogical and petrophysical methods on core samples collected from shallow exploration drilling and outcrop.

The spatial distribution of reservoir properties of deltaic sedimentary rocks in the "Atled Creek" concession in the onshore Niger Delta Basin has been highlighted in the 3-D reservoir model. Gamma ray and density well logs were used to characterize the lithology, whereas neutron porosity and density logs were used to evaluate the petrophysical properties. Depositional facies distributions are modelled by a stochastic object-based approach, which honors the heterogeneity of the reservoir units. The corresponding trend maps (distance to object, maximum curvature, and depth trends) were inputs in constraining the distribution of petrophysical properties within the model.

Four reservoir units that exhibit good to excellent reservoir properties were identified within the concession. The sediments deposited in distributary channels show the best reservoir quality. This dominates the facies assemblages of the deepest studied reservoir unit (EX 2.0), which is the best and only oil-bearing reservoir in the field. The remaining three reservoir units (DX 6.0, DX 7.0 and DX 9.0) are predominantly gas-bearing. Uncertainty and sensitivity analysis helps to assess the obtained volume of hydrocarbons and identify the sensitivity of the model geological parameters. The created model is most sensitive towards uncertainty in the tortuosity factor, sorting and depositional facies of the geological input parameters. The study of the factors controlling the reservoir volume showed that the diagenetically controlled tortuosity factor has the greatest influence on the producible reservoir volume within the scope of its uncertainties.

The examined Cretaceous to Paleogene, continental to marine deposits in the Eastern Dahomey Basin display a high heterogeneity in their lithological assemblages. This variation in lithofacies distributions containing fluvial, estuarine, and shallow marine sandstones and mudstones, as well as marine limestones, accounts for the heterogeneous properties of reservoir rocks.

Reservoir characterization shows a detrital composition dominated by quartz, with subordinate contents of feldspars and rock fragments. The authigenic composition is dominated by pore-filling carbonate cement, and iron oxide (FeOx) and iron oxyhydroxide (FeO(OH)) cements, with subordinate kaolinite, and glauconite grains. Ferroan calcite and ferroan dolomite cements are only present in the sandstone samples from shallow exploration wells in the central to western parts of the basin while siderite cementation occurs in some samples of one formation in the eastern part of the basin. Intense carbonate cementation is absent in the outcrop samples. However, iron oxides (FeOx and FeO(OH)) are only abundant in the outcrop analogs, especially in the Benin Flank east of the Dahomey Basin. Detrital clay-matrix (fine silt to clay) occludes the pores in some of the samples from the shallow wells and outcrops. Solid hydrocarbons also fill the pore spaces in the oil sands of one formation.

The best reservoir qualities are obtained for the arenitic sandstone with a porosity up to 47% and permeability >10,000 mD. The permeability in the matrix-supported sandstone intervals of the well is low. Also, intense FeOx and FeO(OH) cementation in the Benin Flank results in low permeability because of the lack of connections between the remaining intergranular porosity. The major controlling factors on the observed reservoir quality are the amount of carbonate cements, iron oxides (FeOx and FeO(OH)) and clay mineral matrix.

In the limestones porosity is moderate where dissolution and microporosity constitute the main pore types, however, the corresponding permeability in the limestone is low. Cementstones exhibit the poorest optical porosity $\leq 0{,}7\%$ while the best reservoir facies in the limestones are the wackestones and packstones.

Kurzfassung

Kontinentale bis flachmarine Ablagerungssysteme enthalten weltweit bedeutende Kohlenwasserstofflagerstätten. Die Heterogenität der Gesteinsschichten stellt jedoch eine Schwierigkeit bei der genauen Abgrenzung und Erschließung von Lagerstätten, sowie den Förderstrategien dar. Ein besseres Verständnis der Eigenschaften von Lagerstätten und ihrer Verteilung und Kontrolle verspricht einen besseren Explorationserfolg.

Das östliche Dahomey- und das Nigerdelta-Becken in Nigeria entstanden während der Trennung der afrikanischen und der südamerikanischen Platte im späten Mesozoikum, nach dem Auseinanderbrechen des Superkontinents Pangäa. Die lokale und regionale Heterogenität der Lithofazies erschwert die genaue Lokalisierung und Vorhersage der Lagerstättenvorkommen im östlichen Dahomey-Becken im Südwesten Nigerias. Im Gegensatz dazu ist das Nigerdelta im Südosten Nigerias ein erdölführendes Sedimentbecken. Das Verständnis der Heterogenität der Lagerstätten, die Kontrolle der Lagerstättenqualität und verbesserte 3-D-Modelle können in dieser Region einen größeren Explorationserfolg ermöglichen. In dieser Studie werden daher die Kontrollfaktoren eines Reservoirs im Rahmen einer statischen 3-D-Modellierung eines Onshore-Feldes im Nigerdelta-Becken anhand seismischer Daten und geophysikalischer Logs untersucht. Es wird gezeigt, wie wichtig die Lagerstättenqualität für das produktive Lagerstättenvolumen ist. Im Folgenden konzentriert sich diese Studie auf die diagenetische Geschichte und die Bewertung der Lagerstättenqualität von Lagerstättengesteinen im östlichen Dahomey-Becken unter Verwendung petrographischer, mineralogischer und petrophysikalischer Methoden an Kernproben, die aus flachen Erkundungsbohrungen und Aufschlüssen gewonnen wurden.

Die räumliche Verteilung der Lagerstätteneigenschaften der deltaischen Sedimentgesteine in der Konzession "Atled Creek" im Onshore-Nigerdelta-Becken wurde in einem 3-D-Lagerstättenmodell dargestellt. Bohrlochgeophysikalische Gammastrahlen- und Dichtemessungen wurden zur Charakterisierung der Lithologie verwendet, während bohrlochgeophysikalische Neutronen-Porositäts- und Dichtemessungen zur Bewertung der petrophysikalischen Eigenschaften eingesetzt wurden. Die Verteilung der Ablagerungsräume wird durch einen stochastischen objektbasierten Ansatz modelliert, der die Heterogenität der Reservoireinheiten berücksichtigt. Die entsprechenden Trendkarten (Abstand zum Objekt, maximale Krümmung und Tiefentrends) dienten zur Eingrenzung der Verteilung der petrophysikalischen Eigenschaften innerhalb des Modells.

In der Konzession wurden vier Lagerstätteneinheiten identifiziert, die gute bis hervorragende Lagerstätteneigenschaften aufweisen. Die in Mündungsarmen abgelagerten Sedimente weisen die beste Lagerstättenqualität auf. Sie dominieren die Fazies der tiefsten untersuchten Lagerstätteneinheit (EX 2.0), die die beste und einzige ölhaltige Lagerstätte in diesem Gebiet ist. Die übrigen drei Lagerstätteneinheiten (DX 6.0, DX 7.0 und DX 9.0) sind überwiegend gasführend. Die Unsicherheits- und Sensitivitätsanalyse hilft bei der Bewertung des ermittelten Kohlenwasserstoffvolumens und der Ermittlung der Empfindlichkeit der geologischen Modellparameter. Das erstellte Modell reagiert am empfindlichsten auf Unsicherheiten in Bezug auf den Tortuositätsfaktor, die Sortierung und die Ablagerungsräume der geologischen Eingangsparameter. Der diagenetisch gesteuerte Tortuositätsfaktor hat im Rahmen seiner Unsicherheiten den größten Einfluss auf das förderbare Reservoirvolumen.

Die untersuchten kreidezeitlichen bis paläogenen, kontinentalen bis marinen Ablagerungen im östlichen Dahomey-Becken weisen eine hohe Heterogenität in ihren lithologischen Zusammensetzung auf. Diese unterschiedliche Verteilung der Lithofazies mit fluvialen, estuarinen und flachmarinen Sand-, Silt- und Tonsteinen sowie marinen Kalksteinen ist für die heterogenen Eigenschaften der Gesteine der Lagerstätte verantwortlich.

Die Charakterisierung der klastischen Lithologien zeigt eine detritische Zusammensetzung, die von Quarz dominiert wird, mit untergeordneten Anteilen von Feldspäten und Gesteinsfragmenten. Die authigene Zusammensetzung wird von porenfüllenden Karbonatzementen und Eisenoxid- (FeOx) und Eisenoxidhydroxid- (FeO(OH)) Zementen dominiert, mit untergeordneten Kaolinit- und Glaukonitkörnern. Eisenhaltige Kalzit- und eisenhaltige Dolomit-Zemente sind nur in den Sandsteinproben aus flachen Erkundungsbohrungen im zentralen und westlichen Teil des Beckens vorhanden, während Siderit-Zementierung in einigen Proben einer Formation im östlichen Teil des Beckens auftritt. In den Aufschlussproben gibt es keine intensive Karbonatzementierung. Eisenoxide (FeOx und FeO(OH)) sind jedoch nur in den Aufschlüssen reichlich vorhanden, insbesondere in der Benin-Flanke östlich des Dahomey-Beckens. Detritale Tonmatrix (feiner Silt bis Ton) verschließt die Poren in einigen der Proben aus den flachen Bohrungen und Aufschlüssen. Zudem treten feste Kohlenwasserstoffe in den Porenräumen in Ölsanden einer Formation auf.

Die besten Reservoirqualitäten werden für den arenitischen Sandstein mit einer Porosität von bis zu 47 % und einer Permeabilität von >10.000 mD erzielt. Die Durchlässigkeit in den matrixgestützten Sandsteinabschnitten der Bohrungen ist gering. Auch die starke FeOx- und FeO(OH)-Zementierung in der Benin-Flanke führt zu einer geringen Durchlässigkeit, da es keine Verbindungen zwischen der verbleibenden intergranularen Porosität gibt. Die wichtigsten Einflussfaktoren auf die beobachtete Lagerstättenqualität sind die Menge an Karbonatzementen, Eisenoxiden (FeOx und FeO(OH)) und die Tonmineralmatrix.

In den Kalksteinen ist die Porosität mäßig, wobei Auflösung und Mikroporosität die wichtigsten Porentypen darstellen, die entsprechende Durchlässigkeit im Kalkstein ist jedoch gering. Cementstones weisen die geringste optische Porosität $\leq$ 0,7 % auf, während die besten Lagerstättenfazies in den Kalksteinen die Wackestones und Packstones sind.

Contents

Declaration of originality

Chapter 2: As the first author during his PhD studies, Olajide J. Adamolekun performed all sample acquisition, reservoir identification, horizon and fault interpretation, reservoir characterization, facies analysis, 3-D static modelling, data analysis and interpretation. He wrote the entire article with inputs from Michael P. Suess, Benjamin Busch, Idongesit Akwaowoh, Oladotun A. Oluwajana, and Christoph Hilgers. The discussions with all co-authors contributed to this manuscript.

Chapter 3: As the first author during his PhD studies, Olajide J. Adamolekun performed all sample acquisition, thin section analysis, point counting, porosity and permeability measurements, data analysis and interpretation. The XRD measurements benefited the assistance of Mathias Schowtzer and Felix Krause and are highly appreciated. He wrote the entire article with inputs from Benjamin Busch, Nickolaas Molenaar, Michael P. Suess and Christoph Hilgers. The discussions with all co-authors contributed to this manuscript.

Chapter 4: As the first author during his PhD studies, Olajide J. Adamolekun performed all sample acquisition, thin section analysis, point counting, data analysis and interpretation. He wrote the entire article with inputs from Benjamin Busch, Oladotun A. Oluwajana, Michael P. Suess and Christoph Hilgers. The discussions with all co-authors contributed to this manuscript.

1 Introduction

1.1 Motivation

The success of any exploration project in gas, oil or geothermal exploration as well as CO_2 and H_2 storage depends on accurate prediction and location of porous and permeable reservoir lithologies. Whether in conventional or unconventional (including tight reservoirs) hydrocarbon exploration, geothermal and aquifer exploration, carbon dioxide (CO_2) and hydrogen (H_2) storage, reservoir quality is one of the most important elements in reservoir characterization. The assessment of reservoir quality is possible through the use of petrophysical, petrographical, sedimentological, and structural methods from rocks and reservoirs as well as the analysis of well logs (e.g. gamma ray, resistivity, density, neutron porosity, and sonic).

Reservoir characterization and modelling often involve the integration of seismic data, geophysical well logs and geological data. An accurately generated three-dimensional (3-D) model is invaluable in any exploration project. The advantage of 3-D modelling methods is the ability to model complex geological structures and parameters from which economic decisions may be made. In order to minimize risk during decision-making, evaluation of the response of a model to uncertain geological parameters important. A model could also be subjected to sensitivity analysis to know the dominant factor affecting the output or performance of a model.

The heterogeneity of sedimentary rocks makes the accurate prediction of reservoir rocks difficult. Both detrital and authigenic components of a sedimentary rock affect the quality of reservoir rocks (Fig. 1.1) (Bloch, 1994; Morad et al., 2010; Rossi et al., 2002). Porosity distribution in rocks varies within the wide limits of composition, transport history, depositional environment, subsequent diagenetic modification, structural influence and tectonics (Bjørlykke, 2015; Busch et al., 2018a; Monsees et al., 2020b). The degree of sorting, grain size distribution, roundness, and cement and matrix contents all play a significant role in porosity evolution of reservoir rocks (Busch et al., 2018b; Monsees et al., 2020a). Facies distributions, which largely depend on the respective depositional environment, can affect the initial sediment composition and subsequent diagenetic alteration by controlling parts of the available reagents (Busch et al., 2018a; Ding et al., 2013; Monsees et al., 2020b). In continental to shallow marine settings, the rapid depositional facies changes resulting from the constant interaction between sediment supply and sea level changes, controlling the accommodation space, affect the lithofacies assemblages. This in turn conditioned the diagenetic pathway and influences the quality of reservoir rocks. The petrophysical properties of a reservoir rock are a direct result of sedimentological and diagenetic properties.

The qualitative and quantitative evaluation of the effect of the detrital and authigenic components (Fig. 1.1) on possible reservoir rocks in these heterogeneous settings is invaluable in any reservoir-related exploration project. The evaluation of facies, petrographic, sedimentologic and petrophysical properties of sedimentary rocks promise accurate prediction of reservoir rocks. This could only be achieved through an integrated approach comprising field observations from outcrop analogues, core analysis, sedimentological, petrographic, mineralogical, and petrophysical methods as well as the analysis of seismic and well log data, and modelling of the subsurface rock properties in 3-D away from well control.

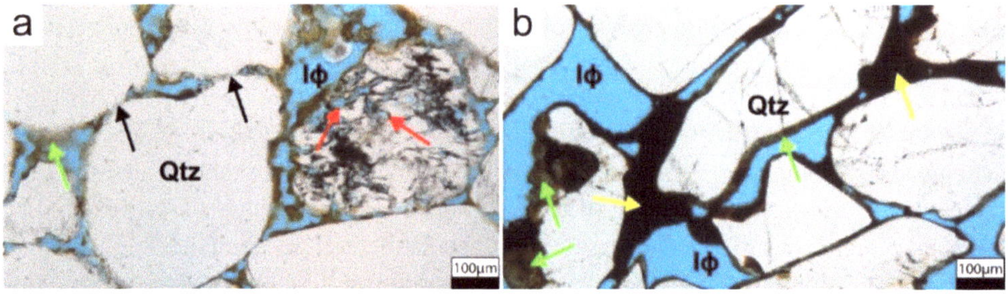

Figure 1.1. Example of detrital and authigenic components. a) Detrital clay matrix (green arrow) filling intergranular pore space (Iϕ) in low compacted (evident by point contact; black arrows) sandstone. Intragranular pore space (red arrows) resulting from the dissolution of unstable detrital grain. b) Detrital matrix (green arrows) and pore-filling FeOx cement (yellow arrows) reduced the initial intergranular pore volume and stabilized the framework.

Coastal sedimentary basins such as the Eastern Dahomey Basin and the Niger Delta in southern Nigeria host gas, oil and solid hydrocarbon (bitumen) accumulations in several fields. However, in contrast to the Niger Delta the success of exploration within the Dahomey Basin is limited. This limited exploration success is a reflection of the facies heterogeneity within the basin and its limited understanding up to present day. The productivity and sustainability of brown fields in the Niger Delta Basin is dependent on the success of the re-interpretation of existing data sets with a view of increasing the reserve base of the region. This involves the generation of three-dimensional (3-D) structural and property models based on new approaches that are readily deployable in field development and management. The most promising pathway to a successful exploration and subsequent field development is the identification of reservoir quality controls in a basin.

1.2 Aim and objectives

The aim of this study is to evaluate the reservoir quality and reservoir quality controls of the continental to shallow marine settings in the Dahomey Basin, and generate a novel three-dimensional static model of deltaic sandstone reservoirs in the "Atled Creek" concession in the Niger Delta Basin. The focus of this research is on:

- Generation of geologically consistent volume-based three-dimensional (3-D) structural model that is valuable in prospect identification.
- The application of an object-based stochastic approach in the generation of three-dimensional (3-D) reservoir models (facies, petrophysical), that is readily deployable for volumetric assessment and reservoir simulation.
- The assessment and evaluation of the distribution of sedimentological, petrographic, and petrophysical properties in siliciclastic and carbonate rocks from continental to shallow marine depositional environments.
- Understanding the controls on the quality of reservoir rocks in continental to shallow marine settings.

1.3 Overview of the thesis

1.3.1 Depositional facies and reservoir modelling (Chapter 2)

The complexity induced by changes in the depositional environment, lithology, and structural and tectonic frameworks often make accurate prospect evaluation difficult. This difficulty can be overcome through a geologically consistent 3-D model, which addresses the temporal and spatial distribution of rock properties within the reservoir units (Abdel-Fattah et al., 2018; Adeoti et al., 2014). A 3-D model allows the evaluation of undrilled areas within a field with the aim of identifying possible prospects and/or increasing the productivity of existing reservoirs.

In this chapter, we discuss the role of an object-based stochastic approach in facies modelling of reservoir units and the corresponding advantage in petrophysical modelling. To capture the heterogeneity of a reservoir and to better constrain the petrophysical properties within a model, a reliable conceptual model must exist on which a more promising reservoir model can be built. A fit-for-purpose 3-D structural and property models is possible through an integrated workflow that utilizes 3-D seismic, well log data and a conceptual model. The volume-based modelling (VBM) workflow in Petrel™ allows the generation of geologically consistent 3-D models based on the inputs from structural, petrophysical, facies, and stratigraphic analyses. Four well log suites complemented by data in regional literature allowed detailed depositional facies and stratigraphic analyses, and the construction of a conceptual model, which aided the facies modelling process. Prospects and leads are identified in the shallow marine onshore Niger Delta Basin by evaluating the property distribution within the generated 3-D models. In a sensitivity study it can be demonstrated that, provided the known uncertainties of the reservoir, the tortuosity factor – a reservoir quality parameter controlled by diagenesis, is the major controlling factor affecting the reservoir volume. The generated 3-D model is deployable in reservoir simulation.

1.3.2 Reservoir quality controls in shallowly buried samples from the Eastern Dahomey Basin (Chapter 3)

Lateral and vertical facies changes are common in transitional to shallow marine environments. The degree of variability often depends on the interaction between land and sea resulting from sea level and sediment supply changes. Lithofacies heterogeneity results in corresponding heterogeneous reservoir properties and complicates the accurate location of good reservoir rocks. This has led to many unsuccessful efforts in hydrocarbon exploration in the Dahomey Basin.

In this chapter, the porosity and permeability distribution in the transitional to shallow marine siliciclastic and carbonate sedimentary rocks of the Eastern Dahomey Basin has been evaluated. Point count data are integrated with the petrophysical data in order to determine the main controls on petrophysical properties. Sandstone permeability is dependent on carbonate cementation and detrital matrix content, with lower contents of both generally resulting in higher permeability. The detrital matrix stabilizes the framcwork in uncemented/weakly consolidated sandstones, but could lead to the production of fines during exploitation. However, the largely kaolinitic (non-swelling clay) composition of the clay mineral matrix poses a low exploitation risk because of the ability of kaolinite to be brought to the well head rather than clogging pore spaces during production. The micropores within the micritic groundmass of the limestone contribute to its porosity and correlates positively with permeability.

1.3.3 Reservoir quality controls of outcrop samples from the Eastern Dahomey Basin (Chapter 4)

The use of outcrop analogs is particularly important in exploring frontier basins where subsurface samples and data are rare. The sedimentological properties controlling the variation in petrophysical properties of rocks are often visible in outcrops. These variabilities are most times at a larger scale than the resolution of bore hole or grain scale measurements (Becker et al., 2017; Schmidt et al., 2020b). However, the exposure of rocks and contact with meteoric water during uplift or shallow burial affects the diagenetic properties of rocks and in turn the reservoir properties. Nevertheless, outcrop analysis is invaluable in the regional identification of key sedimentological elements that may affect reservoir properties. Results are correlated to subsurface samples from chapter 3 and the applicability of outcrop samples to the subsurface will be evaluated to better understand the porosity evolution in the basin.

This chapter provides an understanding of the sedimentological, petrographical and porosity evolution in the basin based on clastic and carbonate rocks collected from 24 outcrop locations. While direct petrophysical measurements are not possible due to the weakly consolidated nature of the samples, quantitative petrographic analysis allows porosity to be estimated by detailed petrographic analysis and point counting.

The observed diagenetic features of the outcrop analogues and shallow subsurface samples differ laterally and vertically. More intense iron oxide (FeOx) and iron oxy-hydroxide (FeO(OH)) cementation are present in outcrops but absent in subsurface samples. Also, the regions in the easternmost part of the basin witnessed more intense FeOx and FeO(OH) cementation compared to the western section. The depositional environments changed vertically from fluvial, through transitional to shallow marine as indicated by conglomeratic sandstones, medium- to coarse-grained sandstone, fine-grained to silty sandstone, shale and claystone, and limestones with marine fossils. The presence of radiaxial calcite in limestones recovered from 2 shallow exploration wells in the westernmost part and poorly developed radiaxial fibrous calcite aggregates in an outcrop in the easternmost part of the basin indicates deposition in restricted water (lake, lagoon) other than shallow marine.

1.4 Parts of this thesis which have been published

Articles

Adamolekun, O.J., Busch, B., Molenaar, N., Suess, M.P. and Hilgers, C., 2022. Petrography and reservoir quality controls in shallow transitional marine Cretaceous-Paleogene deposits in the Dahomey Basin, Nigeria. Journal of African Earth Sciences, Vol. 186, p. 104437. https://doi.org/10.1016/j.jafrearsci.2021.104437

Adamolekun, O.J., Busch, B., Oluwajana, O.A., Suess, M.P., and Hilgers, C., 2023. Diagenetic evolution of continental to shallow marine Mesozoic–Cenozoic deposits, Eastern Dahomey Basin, Nigeria: Implications for reservoir quality. Journal of Applied and Regional Geology. https://doi.org/10.1127/zdgg/2023/0378

Submitted manuscripts

Adamolekun, O.J., Suess, M.P., Busch, B., Akwaowoh, I., Oluwajana, O.A., Hilgers, C. Depositional facies and 3-dimensional modelling of deltaic sandstone reservoirs - an onshore Niger Delta (Nigeria) case study.

Conference contributions

Adamolekun, O.J., Busch, B., Suess, M.P., Akwaowoh, I. and Hilgers, C., 2022. 3D seismic interpretation and reservoir modelling of deltaic sandstone in the onshore Niger Delta Basin, Nigeria. DGMK/ÖGEW Frühjahrstagung, Celle, Germany.

Adamolekun, O.J., Busch, B., Suess, M.P., Akwaowoh, I. and Hilgers, C., 2021. Prospect identification and development in the "Atled Creek" concession, Onshore Niger Delta: Indications from 3D geometrical modelling. Soc. of Pet. Eng., Student Technical Congress. SPE Student Technical Congress, Karlsruhe (online).

Adamolekun, O.J., Busch, B., Suess, M.P., Idongesit Akwaowoh and Hilgers, C., 2021. Structural modelling of Agbada (Tertiary) sandstone reservoirs in "Atled Creek", Onshore Niger Delta, Nigeria. Annual Conference of the German Geological Society (DGGV), Karlsruhe (online), Germany.

Adamolekun, O.J., Busch, B., Suess, M.P. and Hilgers, C., 2021. Diagenesis and reservoir quality of sedimentary rocks of the eastern Dahomey Basin, Nigeria. DGMK/ÖGEW Frühjahrstagung, Celle (online).

2 Depositional facies and 3-dimensional modelling of deltaic sandstone reservoirs - an onshore Niger Delta (Nigeria) case study

2.1 Abstract

Highlighting the temporal and spatial distribution of structural, facies, and petrophysical properties of reservoir rocks is possible with the preparation of three-dimensional (3-D) reservoir models. This is of great significance in the overall strategy for reservoir development, simulation, prediction, and well placement. It also provides a detailed look into the local development of a sedimentary basin. Here we present a modern reservoir modelling workflow applied to a conventional oil field under production in Nigeria.

A stochastic 3-D structural and property reservoir model of the "Atled Creek" concession was prepared from available seismic data and wireline logs to evaluate the reservoir quality and distribution for further field development, reservoir optimization, and expansion. Four deltaic sandstone reservoir intervals, prevalent in the Miocene of the Niger Delta Basin, have been identified and modelled in this study to highlight their general characteristics and reservoir properties. The structural model revealed the dominance of generally W–E trending growth faults, dipping south. A major fault trending WNW–ESE (F1) separates the studied area into two major blocks (north and south) with the existing wells located in the southern block. Fault-related anticlinal structures are the main trap type revealed by the structural model.

The deltaic reservoir units in the Miocene Agbada Formation of the Niger Delta show good to excellent reservoir properties with effective porosity up to 38.2% with an average of 16.7%, and permeability reaches up to 1676.5 mD (avg. 114.7 mD). The interpreted distributary channel facies show the best average porosity of 19.2%. Both oil and gas are present in reservoir unit EX 2.0, a Middle Miocene fluvio-deltaic reservoir, while the remaining reservoirs are entirely gas-filled.

The 3-D static model and the object-based (stochastic) facies model help to constrain our petrophysical property model and could be used in initiating a new development strategy and in reservoir simulation. It helps in better understanding reservoir complexity and in mapping out prospective sites for future expansion and field development. Using an uncertainty workflow, we investigate the main uncertain components of our model, including determination of permeability.

2.2 Introduction

The paralic nature of deltaic reservoirs such as that of the Niger Delta in Nigeria and the complexities in structural style often leads to great heterogeneity in reservoir properties and the structural framework. In this paper, we discuss the properties of four reservoir units within the Middle Miocene of the onshore Niger Delta in the "Atled Creek" concession (the name has been changed due to confidentiality of some provided information) and demonstrate the development of 3-D static models including facies and petrophysical properties for the field. This is achieved with the integration of 3-D seismic data, well log data, biofacies data, and field reports, and highlights the utilization of reservoir models to understand reservoir quality, facies heterogeneity and distribution. The hydrocarbon in this basin has been generated from the deep marine shale of the Akata Formation and migrated up-dip into the overlying sandstone of the paralic Agbada Formation through the faults. The major trapping mechanisms in the basin are the roll-over anticline and fault assisted clossures. The shale interbeds within the Agbada Formation forms viable top and bottom seals for the hydrocarbon.

Facies heterogeneity imposed by the complexity of the source area input, depositional environment, compartmentalization, structural framework, geological, and tectonic setting amongst others poses a difficulty in accurate prospect delineation, field development, and production strategies. The performance of a well and the overall success of a field development are dependent on the distribution of its reservoir parameters in 3-D (Alo & Ehinola, 2018; Haque et al., 2016; Okoli et al., 2021; Radwan et al., 2021; Ringrose & Bentley, 2015). Understanding the facies architecture and the evaluation of key petrophysical, sedimentological and/or depositional properties are of significance in reservoir characterization and modelling (Abdel-Fattah et al., 2018; Alo & Ehinola, 2018; Lambert-Aikhionbare, 1981; Magbagbeola & Olayinka, 2019; Ringrose & Bentley, 2015; Weber, 1971).

The complexity of small-scale (up to cm-scale) reservoir characterization is further related to the scarcity of core samples, especially in weakly consolidated lithologies present in the Niger Delta. However, geophysical data has proven invaluable in subsurface reservoir evaluation for gas, oil and geothermal exploration as well as CO_2 and energy storage such as hydrogen. Reservoir characterization, a process that is aimed at the qualification and quantification of reservoir rocks and fluid properties such as porosity, permeability, shale/clay volume, and fluid saturation amongst others attempt to group reservoir rocks into different hierarchical units (Adelu et al., 2019; Alo & Ehinola, 2018; Islam et al., 2020). Porosity and permeability may generally be directly proportional within reservoir facies and yet have both low and high permeability zones within a reservoir unit of equal porosity values but different reservoir facies in a given reservoir. These different porosity-permeability relationships are expressions of different hydraulic units and responses to changes in sedimentological and diagenetic properties of rocks (Dezfoolian et al., 2013; Peralta, 2009; Ringrose & Bentley, 2015; Wimmers & Koehrer, 2014).

Usually, to overcome the complexity of field development and production, several models are generated during the lifetime of a field using conventional seismic and well log data. These models attempt to reconstruct the structural framework and identify the property distribution within the zone of interest to address specific reservoir development challenges. An improved understanding of the temporal and spatial distribution and controls on reservoir properties within a field is necessary for reliable volumetric analysis and reservoir development. Data from seismic reflection surveys and wireline logging are important assets in developing regionally consistent 3-D models. The analysis/interpretation of seismic data allows a quantitative division of the reservoir models into layers and cells in which attributes are allocated based on the geological properties of the model. The analysis and interpretation of property distribution from the wellbore is critical to the accurate construction of representative geological models for effective reservoir management. Key economic, reservoir management and development decisions are made based on the result of volumetric analysis. Central to the computation of reservoir volume is a reliable static model which often comes with some uncertainty in the input parameters (facies, petrophysics, contact, structure, velocity). The evaluation/quantification of the uncertainty and the sensitivity of a model to relevant geological parameters is important in decision making.

2.3 The Niger Delta Basin

The hydrocarbon-bearing Niger Delta Basin lies within the Gulf of Guinea and extends over 70,000 km^2 (30,000 mi^2) forming the youngest and the largest sedimentary body that follows the separation of the African and South American plates in the South Atlantic (Fig. 2.1; Doust & Omatsola, 1990; George et al., 2019; Magbagbeola & Willis, 2007). The extensional tectonic events from the Cretaceous to Palaeocene have influenced the evolution and sedimentation in the Niger Delta and the neighboring Dahomey Basin (Adamolekun et al., subm.; Adamolekun et al., 2022; Lambert-Aikhionbare, 1981). Regionally, sedimentation was largely controlled by marine transgressions and regressions related to eustatic sea-level cycles with varying duration (Doust & Omatsola, 1990; Reijers, 2011).

The fill of the Niger Delta Basin is largely progradational with depositional complexities arising from a series of fault-controlled subbasins generating depobelts striking mostly NW-SE subparallel to the present shoreline (Fig. 2.1; Magbagbeola & Willis, 2007).

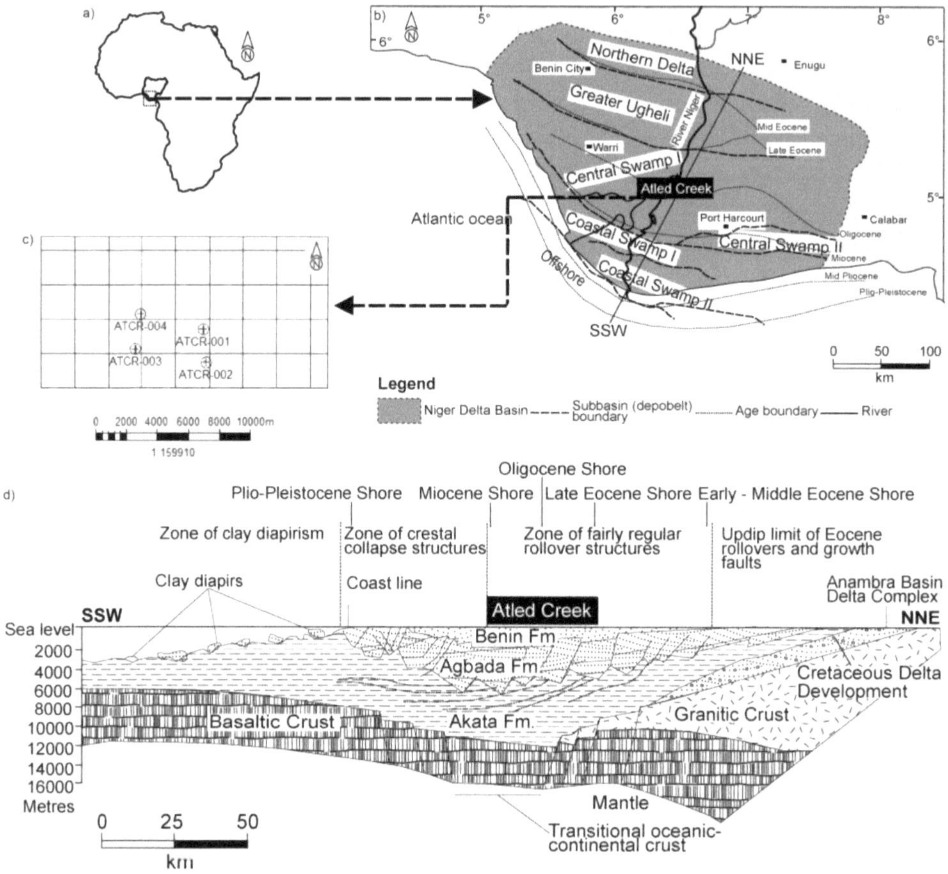

Figure 2.1. Map of Africa showing the location of the Niger Delta (dashed box) along the coast of Nigeria. (b) Map of Southern Nigeria showing the location of the "Atled Creek" concession within the Central Swamp subbasin (depobelt) of the Niger Delta. The structurally defined sub-basins become progressively younger from onshore to offshore (after Doust and Omatsola, 1990; Short and Stäuble, 1967). c) The "Atled Creek" concession map showing the seismic survey coverage (inline, crossline) and well locations (d) Cross-section of the Niger Delta Complex from offshore (SSW) to onshore (NNE) showing the Benin, Agbada and Akata Formations and the structural framework of the basin (after Short and Stäuble, 1967).

Thus, the development of the delta proceeds in discrete structural controlled stratigraphic units that fill in the subbasins. Furthermore, under increasing deltaic sediment loading, the underlying overpressured shale is forced to squeeze upward and basinward on the slope (Knox & Omatsola, 1989) of the Nigerian Shelf. A corresponding basinward structural collapse of the shelf resulted in normal faults (striking NW-SE) and created accommodation space for additional substructures with shallow-marine fluvio-deltaic sediments.

More so, subsidence was slowed by the depletion of mobile shale under the area of a subbasin, shifting sedimentation basinward. A new subbasin is developed seaward of the previous one by declining accommodation which forced the sediments to prograde. Extensional, syndepositional growth faults and crestal extensional faults, both striking northwest-southeast on rollover anticlines are major structural features of the basin and are of great significance to the trapping system (Adeigbe & Alo, 2017; Lambert-Aikhionbare, 1981; Magbagbeola & Willis, 2007).

The stratigraphy of the Niger Delta is made up of three diachronous, widely recognized units of siliciclastic rocks which constitute a succession of generally regressive, offlapping sediments. These are from oldest to youngest: Akata, Agbada, and Benin Formations (Fig. 2.2; Doust, 1990; Knox & Omatsola, 1989; Reijers, 2011; Short & Stäuble, 1967; Weber & Daukoru, 1975). Boundaries between these formations are diachronous and comprise broadly defined regional clinoforms (younging basinward) of the continental margin (Figs. 2.1, 2.2; Magbagbeola & Willis, 2007).

The Akata Formation consists of a thick unit of medium to dark grey, fairly hard, offshore/deep marine, prodeltatic shales (Fig. 2.2) deposited from Early Cretaceous onward on the Nigerian shelf and continental margin. In places, it can also be soft, sandy or silty (Doust, 1990; Reijers, 2011). The shale is under-compacted and as a result, in many instances contains lenses of abnormally high-pressured siltstone and/or fine-grained sandstone (Lambert-Aikhionbare, 1981; Short & Stäuble, 1967). The base of this formation has not been drilled but the maximum thickness is estimated to be about 7000 m and the known age ranges from Cretaceous to recent (Lambert-Aikhionbare, 1981; Merki, 1970; Short & Stäuble, 1967; Tuttle et al., 1999).

The paralic Agbada Formation (Fig. 2.2) consists of an alternating sequence of medium- to fine-grained sandstones, siltstones, light to dark grey shales (fairly consolidated), and silty shales. These are considered to be deposited in the delta front, distributary channel, and delta plain systems (Lambert-Aikhionbare, 1981; Tuttle et al., 1999; Weber & Daukoru, 1975) downlapping on the Akata Formation. The dominant minerals in the sandstones are quartz, and Na-feldspar, with a minor (subordinate) amount of K-feldspar, kaolinite, and illite. Furthermore, the sandstones are sometimes locally calcareous, glauconitic, and shelly while the occurrence of glauconite is also locally recorded in the shale (which contains mainly kaolinite with a small amount of illite and montmorillonite mixed layer clays). The formation may attain a maximum thickness of 3,940 m and thins towards the northwestern and eastern margins of the delta (Lambert-Aikhionbare, 1981; Short & Stäuble, 1967). This formation is the target of all hydrocarbon exploration in the basin as it contains nearly all the hydrocarbon accumulations.

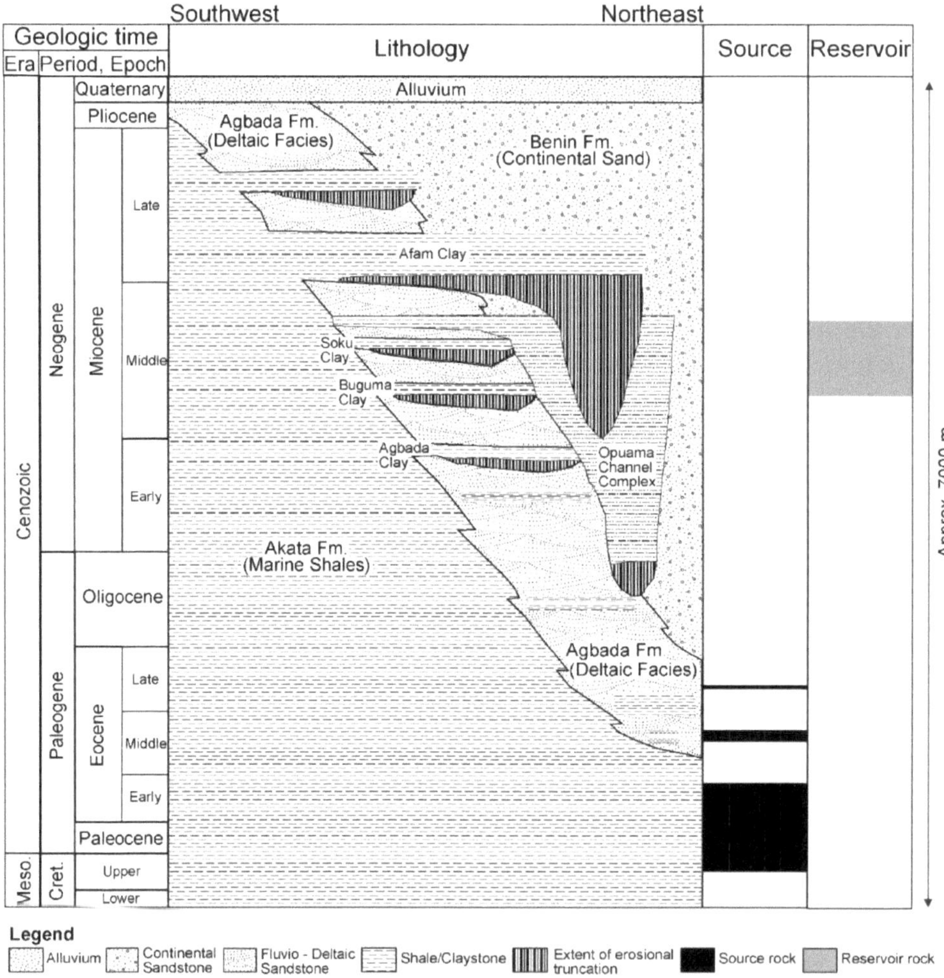

Figure 2.2. Regional stratigraphic column of the Niger Delta showing the interactions between the Akata, Agbada and Benin Formations (modified from Doust and Omatsola, 1990).

The age of the Agbada Formation decreases from north to south of the basin and varies from Eocene to recent. Internally, the Agbada Formation is subdivided by the Agbada, Buguma, Soku and Afam Clays deposited in several canyons. The Opuama Channel Complex covers about 1300 km^2 in the north-western part of the basin and is thought to have begun as a submarine canyon (incision). This truncated the paleodelta surface during the Upper Eocene/Early Oligocene sea-level lowstand and resulted in the development of a network of deep and extensive channels in the Upper Eocene to the Early Middle Miocene of the Niger Delta Basin (Magbagbeola, 2005; Petters, 1984; Udo & Ekweozor, 1990). The canyon extends offshore for at least 30 km and the associated truncations/channels serve as pathways for downslope debris flow to deeper marine environments to form turbidites and submarine fan deposits (Petters, 1984; Udo & Ekweozor, 1990). More so, during marine transgressions, pelagic clays are also dumped in the channels (Udo & Ekweozor, 1990).

The Benin Formation (Fig. 2.2) predominantly consists of massive, highly porous and fresh water-bearing continental sandstones (70 to 100% of the formation) with thin interbeds of shales (usually containing some plant remains and dispersed lignite) which are considered to be deposited in braided stream systems. Quartz, K-feldspar, and minor amounts of Na-feldspar dominate the mineralogical assemblage of this formation. The formation can attain a thickness of 1,970 m and the base is most often arbitrarily defined by the deepest fresh water-bearing sandstone characterized by high resistivity (easily identified in resistivity log analyses) or the appearance of the first marine foraminifera (Lambert-Aikhionbare, 1981; Short & Stäuble, 1967). The Benin Formation generally ranges in age from Oligocene to recent.

2.4 Dataset and Methods

The methodology (Figure 2.3, sections 2.4.1 – 2.4.5) employed in the depositional facies and 3-D static modelling of the "Atled Creek" concession helps in discussing the prospectivity of the field considering the concept of reservoir geometry, structural framework, facies, and petrophysical properties. The reservoir model is based on seismic and well interpretation, coupled with conceptional ideas derived from general facies models and regional information available in confidential reports, and published works.

2.4.1 Dataset

The seismic survey and data from production wells used for this research are located within the central swamp subbasin (depobelt; Fig. 2.1 b), onshore Niger Delta. These include a 186 km^2 pre-stacked depth migrated (PSDM) 3-D seismic volume, a suite of wireline logs from four production wells (ATCR-001, ATCR-002, ATCR-003, ATCR-004), containing gamma-ray (GR), spontaneous potential (SP), resistivity (RT), neutron (NPHI), density (RHOB) and sonic (DT) logs. Other available data are deviation surveys, check-shot data, formation tops, and biofacies data. Additionally, lithological information is obtained from published reports and literature (Akaegbobi et al., 2003; Lambert-Aikhionbare, 1981; Lambert-Aikhionbare, 1982; Lambert-Aikhionbare & Shaw, 1982; NNPC-Report, 1997 to 2020; Oboh, 1992a, 1993, 1995; Odedede, 2019). Depth-based evaluation of the published petrographic, sedimentological, and petrophysical data allowed the creation of an internally consistent database used in this study. This allows the empirical definition of the distribution of processes and properties which are important in addressing the objectives of this study. All modelling in this study has been performed using Petrel™ E&P software.

2.4.2 Conceptual model

A reliable conceptual model was generated based on the available seismic and well log data, biofacies data, published regional literature and other confidential information. This involves the structural, depositional, sedimentary facies and diagenetic models and scenario that characterizes the gross depositional environment and strata distribution in the field. The resulting 3-D conceptual model allows the description of physical features of the geologic system of the Niger Delta Basin. The essential geological scenario from tectonics, sedimentation, stratigraphy, burial, hydrocarbon generation, migration and accumulation are considered.

2.4.3 Seismic and well data interpretation

The available georeferenced seismic and wireline log data were loaded into Petrel™ E&P software. Quality checking (QC) was performed on the provided data by checking the survey parameters (inline, crossline, frequency), seismic, and well log quality. The well tops were imported and edited where necessary based on the gamma-ray (GR) and resistivity (RT) log signature, and the reservoir units of interest were identified and named (DX 6.0, DX 7.0, DX 9.0, EX 2.0; all within the Miocene sedimentary strata of the Agbada Formation). A shale baseline of 100 API (0 to 80 API = sand, 80 to 100 API = silt, >100 API = shale) was used for lithological identification.

Depositional facies were interpreted based on the typical GR signature for sedimentary depositional environments (Fig. 2.4) (Cant, 1992; Nazeer et al., 2016; Radwan, 2020). The regional geological knowledge of the basin and field provides a valuable asset in interpreting the depositional facies based on well log (electrofacies).

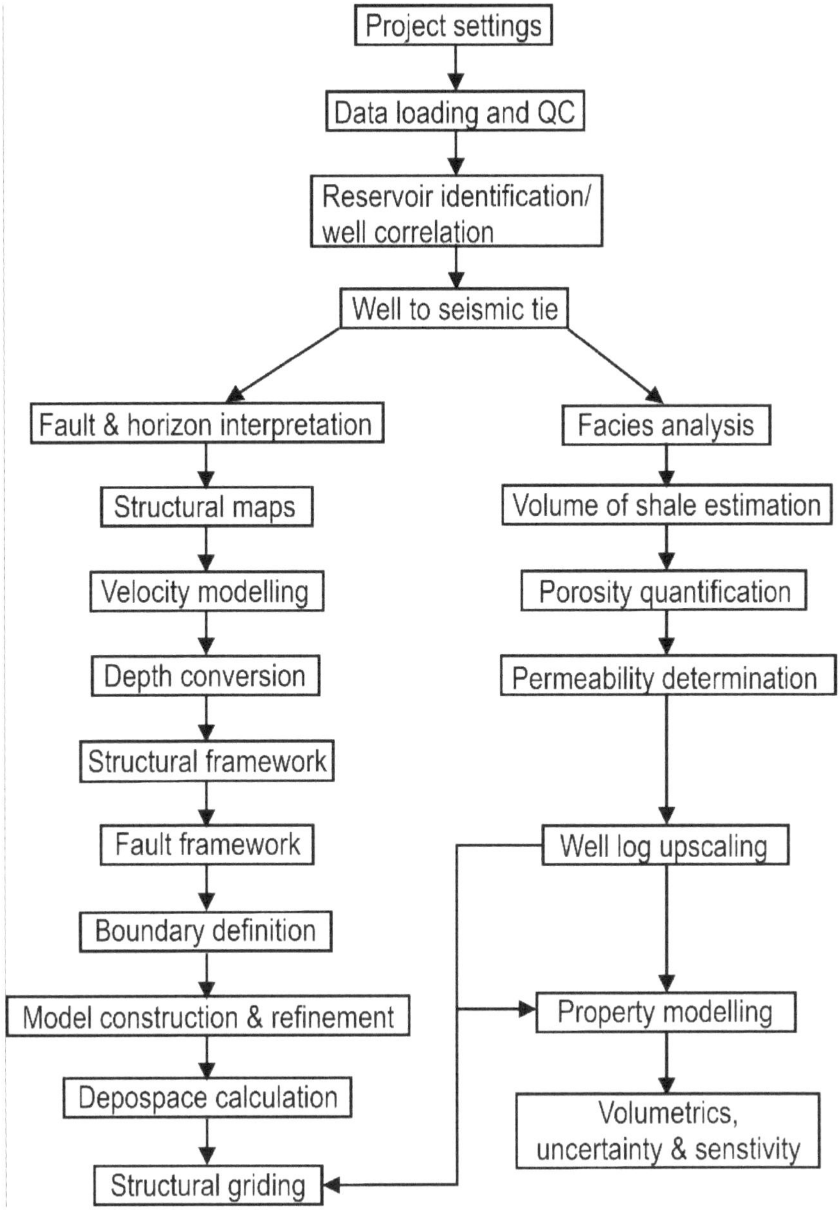

Figure 2.3. Workflow for the depositional facies and reservoir property modelling of the "Atled Creek" concession. (modified from Doust and Omatsola, 1990).

The well log signatures/shapes are indicators of grain size distribution and thus reflect the depositional environment of these rocks (Fig. 2.4).

GR log pattern	Cylindrical/box	Funnel	Bell	Symmetrical	Serrated/irregular
Typical GR trend	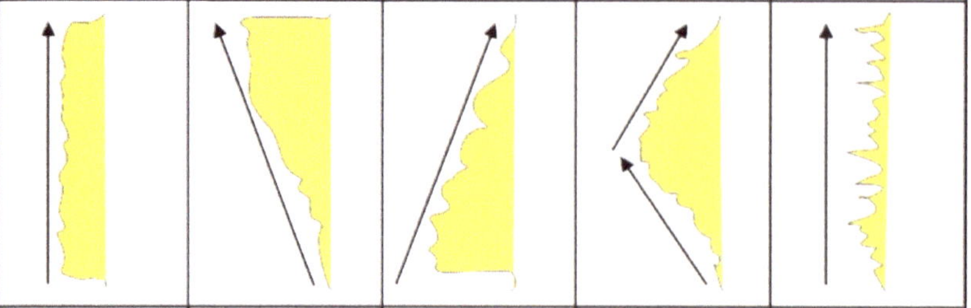				
Sediment supply	Aggrading	Prograding	Retrograding	Prograding & Retrograding	Aggrading
Grain size	Relatively consistent	Increases upward	Decreases upward	Coarsening upward trend change to finning upward sequence at the top	Intermixed shale and sand
Characteristics	Sharp top and base with consistent trend	Cleaning upward trend with abrupt top	Dirtying upward trend with abrupt base	Ideally rounded top and base	Irregular patterns/ spikes
Common depositional environment	Fluvial channel, prograding delta distributaries, carbonate shelf, reef, submarine canyon fill, Aeolian dunes, evaporite fill	Crevasse splay, river mouth bar, delta front, shoreface, submarine fan lobe	Fluvial point bar, tidal point bar, deltaic distributaries, proximal deep sea	Reworked offshore bar, transgressive shelf sands, mixed tidal flats	Fluvial flood plain, mixed tidal flat, debris flow, storm dominated shelf, canyon fill, deep marine slope

Figure 2.4. Typical gamma ray (GR) signature and sedimentological characteristics for different depositional facies (Cant, 1992; Nazeer et al., 2016; Radwan, 2020).

The well to seismic tie-in was done by generating a synthetic seismogram using the checkshot survey, calibrated sonic log (DT), density log (RHOB), acoustic impedance (AI), reflection coefficient (RC), and deterministic wavelets (Extended White; Supplementary material 2.1). This is necessary because the well data are available in the depth domain (m) while the seismic data is in the time domain (ms). A bulk shift of 10 ms was necessary to achieve a fairly good tie between the well and the seismic (Supplementary material 2.1).

Prior to the fault interpretation, structural smoothening, ant tracking and variance attribute analyses were performed on the seismic volume with the former serving as input for the latter in each case (Supplementary material 2.2). These attributes provided valuable visual guidance during the manual fault interpretation. Horizon and fault interpretations were performed within the entire 3-D cube consisting of 852 in-lines and 441 crosslines (Fig. 2.1 c; Supplementary material 2.2 c). A total of 26 faults (F1 to F26) represented by reflection/structural discontinuity were identified and mapped (Supplementary material 2.2 c). Formation tops of interest were displayed on the seismic and the corresponding reflections were mapped as horizons (DX 6.0, DX 7.0, DX 9.0, and EX 2.0; Supplementary material 2.2 c). A boundary polygon was created from the horizon for the subsequent generation of maps (surfaces). The interpreted horizons were used to generate time structure maps (Supplementary material 2.3).

Conventional petrophysical analysis was performed on the wireline logs using suitable formulas (Supplementary material 2.4) and the Petrel™ calculator to generate the petrophysical properties of the reservoirs. Permeability was calculated from porosity in the model using the function in equation 2.1 (Onengiyeofori et al., 2019).

$$k = 10{G_d}^2 \times \phi^{3.64+m} * {S_i}^{-3.64} \tag{2.1}$$

Where k is permeability, ϕ is porosity, G_d is the dominant grain size, m is cementation exponent, S_i is the sorting index of the formation (varying from 0.7 in the extremely well-sorted sample to 1 in the poorly sorted sample). All these parameters were deemed to be uncertain and were investigated in the frame of a sensitivity study. The facies log was created based on the gamma ray and density log signatures, and the available literature and reports.

2.4.4 Velocity modelling and depth conversion

A velocity model was generated using the average interval velocity in a 4-layer case. For this purpose, four additional horizons were mapped above the horizons of interest (reservoir units), and the corresponding time maps were generated. This is necessary to capture the velocity changes downhole, because the thickness of the zone of interest may not be large enough to give a reasonable trend in velocity with depth. The time–depth relationship (TDR) from the checkshot data allows a quick look into the velocity variation (supplementary material 2.5). Based on the observed TDR, the mapped horizons (reservoir units and the additional horizons) were grouped and formed a 4-layer case for the velocity model. The elevation reference at the top of the survey (sea level) was set to zero and interval velocity for subsequent layers was calculated. These were used in the interval velocity modelling and the well top correction was applied. The generated velocity model was used for the depth conversion of our interpretations through the domain conversion process in Petrel™ software.

2.4.5 Reservoir modelling

Based on the conceptual model of the field and after data interpretation, we proceeded to the reservoir modelling processes. Here, the interpreted depth grids together with the fault sticks and derived reservoir properties (facies, porosity, permeability, water saturation) served as major inputs in the model. The ideas from the conceptual model provided additional inputs and guided the modelling processes.

In the creation of the structural frameworks that define the reservoir geometry, we used the volume-based model (VBM) algorithm in Petrel™ (by inputting the results of the seismic and well log interpretation) to build a geologically consistent 3-D structural model. Structural gridding was done to create stair-step-faults to minimize edge squeezing of the model skeleton. In the frame of the 3-D gridding process, cells with almost identical cubic geometry and volume were created within the defined space, with an area resolution of 100 m by 100 m. Following the establishment of the fault-horizon relationship, 3-D pillars along the fault framework model were created. This was done by respecting the law of superposition and cross-cutting relationships (Haque et al., 2016).

The generated property logs (values of the log property) were upscaled to the resolution of the 3-D grid using arithmetic operations. Values at wells are exact and the corresponding uncertainty is assumed to be zero since absolute information is present at well locations. This upscaling process attempts to reduce a large number of fine-scale grid blocks of a geological model to a smaller number of coarse-scale grid blocks while retaining the underlying geological properties (Milad et al., 2020).

During the data analysis, and from the upscaled depositional facies properties, vertical facies proportion curves were generated for the pro-delta, channel, mouth bar, delta plain and delta front facies log based on well log interpretation. The upscaled facies log was used to generate the facies model in an object-based stochastic approach (object modelling) to reveal the variability in each reservoir. The pro-delta was set as background in the model, while object bodies that reflect the other depositional facies, constrained by the well observation, were distributed randomly in the volume. Information from the paleogeography, regional geology and the conceptual model guided the object bodies dimensions and geometry. Generated vertical proportion curves were used as body insertion probability trends for the corresponding facies. The facies model was set to honor at first the facies proportion, rather than being object dependent, because the generated facies proportion curves are based on the well log data and deemed to be more representative for the reservoir in the reference case. To inform the property distribution about depositional trends we use geometric information of the facies object, obtained during the facies modelling step. Available properties are the depth trend, object distance, and curvature (volume trends) of the objects. The depth trend created a property that shows the vertical distance of a cell's center to the edge of the body (vertical; top and base of the body), the object distance created a property that contains the distance to the body's edge (lateral; side to side of the body) and the object curvature created a property that contains the curvature (inverse radius) of the object's edge or its extension inside and outside the body.

The obtained characteristic volume trends from facies modelling (depth, object distance, object curvature) were used in the co-kriging process (as secondary variables) to constrain the distribution of the petrophysical properties. Upscaled porosity and permeability logs were stochastically distributed throughout each defined layer in each reservoir unit. The Gaussian random function simulation algorithm in Petrel™ (GRFS) allows the propagation of the upscaled well data with the co-kriging option for an analyzed distribution by normal scoring into the 3-D grid (Webster & Oliver, 2007). For each property obtained, a variogram was previously developed using the data analysis tool in Petrel™. During the petrophysical modelling, we use the variogram and transformation from data analysis.

Fluid contacts were created and the corresponding contact maps were generated. The water saturation modelling involved the generation of J-function (Eq. 2.2 - 2.3; Leverett, 1941) as;

$$J_f = [P_c / \{\cos\theta \times \sigma\}] \times \sqrt{\frac{3D\ k}{3D\ \phi}} \tag{2.2}$$

$$P_c = h \times [\{(\rho_1 - \rho_2) \times FWL \times a\} / 100000] \tag{2.3}$$

Where J_f = J-function, P_c = capillary pressure, Θ = contact angle (degree), σ = interfacial tension (m/t^2, dynes/cm), 3D k = model permeability, 3D ϕ = model porosity, h = height, ρ = density (wetting and non-wetting phases), FWL = free water level, and a = acceleration due to gravity. The FWL is defined as the depth where the capillary pressure between water (denser) and hydrocarbon (less dense) phases is zero. This is taken as the height above contact (the lowest level where water saturation approaches 1) generated using the geometrical modelling process in Petrel™.

The J-function was then applied to calculate the water saturation (S_w) of the reservoir units (Eq. 2.4) and the irreducible water saturation (Swirr) was used to normalize the water saturation (Eq. 2.5).

$$S_{w_j} = P_c\,(S_w) \big/ \sigma\cos\theta \sqrt{\frac{3D\ k}{3D\ \phi}} \tag{2.4}$$

$$S_{w_j} = \emptyset / (J_f + S_w) + S_{wirr} \tag{2.5}$$

2.4.6 Reserve volumetric estimation

The volumetric analysis allows the quantification of the volume of hydrocarbon present in the reservoir. From the generated static model, reserve estimation in terms of stock tank oil initially in place (STOIIP) and gas initially in place (GIIP) for the reference model were calculated (Eq. 2.6, 2.7).

$$STOIIP = \frac{[7758 \times A \times h \times \phi \times (1\text{-}S_w)]}{B_o} \tag{2.6}$$

$$GIIP = \frac{[43{,}560 \times A \times h \times \phi \times (1\text{-}S_w)]}{B_g} \tag{2.7}$$

Where A = area (acres), h = net pay thickness (ft), ϕ = porosity, S_w = water saturation, B_o = oil formation volume factor (= 1.5), B_g = gas formation volume factor (= 0.0028).

2.4.7 Sensitivity analysis and uncertainty quantification

After a reference or base-case model was generated we aimed to understand the uncertainty and sensitivity of producible hydrocarbons present in the reservoir. Uncertainty analysis was done based on preset cases (low, mid and high), which are dependent on preset simulation realization values. Also, the sensitivity of the volumetric calculation (STOIIP, GIIP) from the static model to uncertain geological parameters was tested. The uncertainty and sensitivity analyses were done based on the workflow developed in Petrel^{TM} and incorporated 3-D grid, facies modelling, fluid contacts, and petrophysical modelling (porosity, permeability, water saturation).

Geological models often come with uncertainties in some parameters which are related to the scarcity of data in space as wells penetrate a narrow vertical section of the studied volume. With a proper uncertainty and sensitivity workflow, the effect of this on the decision-making process can be minimized by estimating different case scenarios for the uncertain variables and the output volumes. During the geological uncertainty quantification, we developed a workflow in which we create realizations of static reservoir properties for uncertain parameters like porosity of facies, facies proportions and reservoir quality parameters like the tortuosity factor, sorting, cementation factor, and the formation factor. These are either directly impacting results (e.g. facies) or are included as components in the calculation of other parameters in the model (e.g. porosity in permeability).

Having built a reliable workflow and preset the values for the uncertain parameters to be tested, the uncertainty quantification was conducted by the Monte-Carlo sampler simulation method using multiple simulation jobs (250 runs) in which values for each realization parameters are randomly and uniquely defined by probability distribution functions. The computation returns values taken from each probability distribution function and corresponding volumes in each run. After the 250th run, a final volume distribution was obtained. We analyzed and evaluate the simulation results to obtain three quantiles; low-case (P10), mid-case (mean; P50) and upper-case (P90) for the model. The importance of the statistical quantification of P10, P50, and P90 is that it reflected the range of STOIIP and GIIP that are valuable in decision-making. They are the base of economic evaluations, calculating the expected monetary value (EMV) and provided an economic valuation model for the alternative cases.

To further study the importance of individual uncertainty parameters for our results, we investigated the sensitivity of the uncertain model parameters (see below, Table 2.1). This was tested using an equal spacing sampler of the uncertainty range (number of samples = 5 per parameter; the number of runs = 40) of main uncertain parameters: Facies proportions (channel and delta plain, delta front, mouth bar), porosity (high, low, base), cementation factor, tortuosity factor and sorting. To understand the impact of the spatial distribution of facies objects we also varied the seed for the initialization of the spatial realization.

A tornado plot of the results of the sensitivity analysis shows the relative influence of the uncertain geological parameters on reservoir volumetrics. This allows the identification of the most sensitive parameters in the model, which can be used to determine measures needed to reduce uncertainty and to improve the basis of an economic decision.

2.5 Results

2.5.1 Reservoir units

Four laterally continuous reservoir units (DX 6.0, DX 7.0, DX 9.0, and EX 2.0) have been identified between 2700 and 4051.85 m in the "Atled Creek" concession (Fig. 2.3, 2.5; Table 2.1). These are within the Miocene sedimentary deposit of the Agbada Formation in the onshore Niger Delta Basin. The thickness of each reservoir unit varies from 28.3 – 47.3 m (DX 6.0, avg. 37 m), 22.3 – 137.8 m (DX 7.0, avg. 63 m), 29.0 – 62.1 m (DX 9.0, avg. 42 m), and 52.1 – 64.9 m (EX 2.0, avg. 61 m; Table 1). The developed model is based on the regional conceptual model of the basin and the interpreted seismic and well log data.

2.5.2 Basic assumptions and the conceptual model

The recognized petroleum system of the Niger Delta Basin is the Akata-Agbada petroleum system in which the marine shales of the Akata Formation are the principal source rocks (Lambert-Aikhionbare, 1981; Tuttle et al., 1999). The generated hydrocarbon migrated up-dip through the normal and growth faults (typical of extensional settings) which constitute the main structural elements in the basin (Fig. 2.1 d). Hydrocarbon accumulation takes place within the deltaic sandstone of the Agbada Formation which constitutes the main reservoir unit in the onshore and shallow sections of the basin.

The pay zones of the reservoir units are mainly made up of channel and mouth bar sediments (Fig. 2.5, 2.6), with additional minor contributions from the sandy to silty parts of the delta plain and delta front. Anticlines, which are associated with mud diapirs and the normal, and growth faults (which are largely syndepositional) are the main trapping elements within the basin. Shale sequences within the Agbada Formation are due to the paralic nature (resulting from a series of transgression and regression) of the delta and constitute the major top and bottom seals. These are often the argillaceous pro-deltaic, delta front and delta plain sediments (Fig. 2.6) deposited during transgression (flooding). The basic assumption in the trend or distribution of sediments is that the coarser sediments (channel and mouth bar) are deposited during the regressive phase (shallowing) while the finer sediments are deposited during marine transgression (deepening). This brings about the observed alternation of fine and coarse sediments in the basin. Also, coarser sediments (due to higher density) are deposited at the points of maximum curvature along the channels (Fig. 2.6) while the straight part of the channel witnessed relatively finer sediments deposition. The mouth bars feed through the channels to which they are attached and become finer seaward.

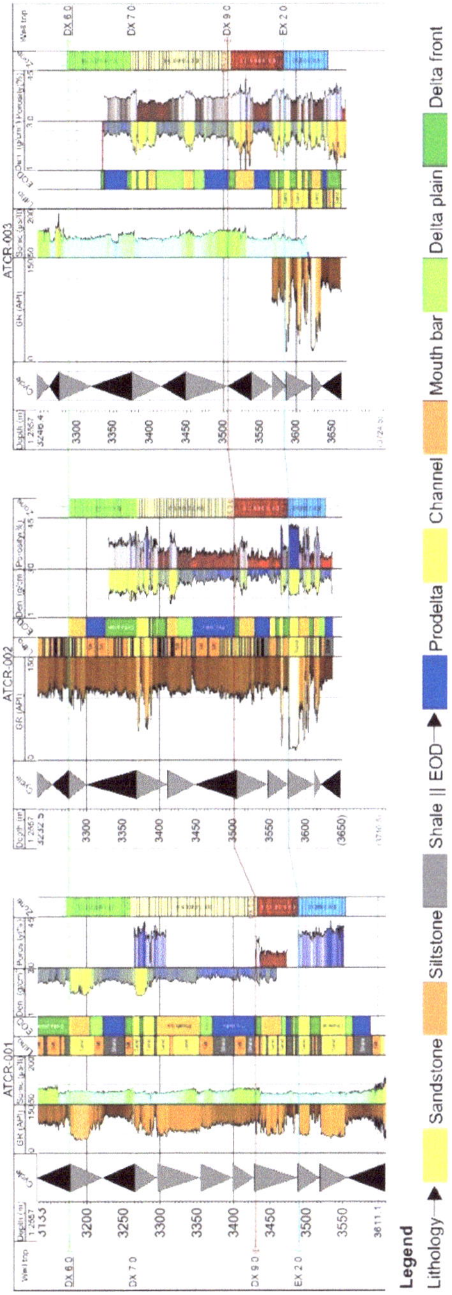

Figure 2.5. Lithofacies and depositional facies interpreted from wireline logs. Shallowing (grey triangle) to deepening (black triangle) lithological sequences are interpreted to represent responses to sea level fluctuations.

Table 2.1. Facies and petrophysical properties of the shallow marine reservoirs in "Atled Creek" concession, Niger Delta from reference wells (a) and (b) uncertainty table of effective porosity estimate per facies based on well statistics investigated in the uncertainty and sensitivity study. Color coding indicates relative high/Low.

(a)	Parameter	Reservoir			
		DX 6.0	DX 7.0	DX 9.0	EX 2.0
Reservoir property	Top (m)	2977.0 - 3347.0 (avg. 3202.75)	3278.4 - 3466.2 (avg. 3360.0)	3445.5 - 3699.4 (avg. 3555.88)	3502.8 - 3784.7 (avg. 3618.38)
	Base (m)	3005.3 - 3386.3 (avg. 3239.4)	3323.8 - 3514.3 (avg. 3423.43)	3488.7 - 3761.5 (avg. 3597.85)	3567.4 - 3849.6 (avg. 3679.25)
	Thickness (m)	28.3 - 47.3 (avg. 36.7)	22.3 - 137.8 (avg. 63.4)	29.0 - 62.1 (avg. 42.0)	52.1 - 64.9 (avg. 60.9)
	Vsh (%)	0 - 70 (avg. 27.6)	0 - 90 (avg. 19.1)	0 - 100 (avg. 28.6)	0 - 100 (avg. 24.9)
	Φ_{tot} (%)	10.9 - 24.7 (avg. 16.6)	6.3 - 33.1 (avg. 22.7)	3.1 - 42.8 (avg. 19.7)	6.4 - 41.7 (avg. 21.8)
	Φ_{eff} (%)	3.3 - 24.7 (avg. 12.7)	9.2 - 29.5 (avg. 16.7)	4.2 - 38.2 (avg. 13.3)	9.0 - 28.0 (avg. 18.2)
	K (mD)	23.3 - 271 (avg. 99.7)	14.4 - 464.6 (avg. 105.3)	1.3 - 1676.5 (avg. 99.4)	13.3 - 396.4 (avg. 126.7)
	F	13.3 - 68.3 (avg. 34.6)	9.3 - 94.3 (avg. 36.8)	3.9 - 460.4 (avg. 71.5)	10.3 - 99.0 (avg. 30.1)
	S_{wirr} (%)	8.2 - 18.5 (avg. 12.8)	6.8 - 21.7 (avg. 13.1)	4.4 - 48.0 (avg. 17.6)	7.2 - 22.2 (avg. 11.8)
	S_w (%)	20 - 100 (avg. 37.9)	20 - 100 (avg. 78.4)	20 - 100 (avg. 58.8)	2.0 - 100 (avg. 36.3)
	S_h (%)	0 - 80 (avg. 62.1)	<0.1 - 80 (avg. 21.6)	<0.1 - 80 (avg. 41.2)	<0.1 - 98.0 (avg. 63.7)
Contact	GWC	3232.26	3294.37	3484.29	-
	GOC	-	-	-	3602.69
	OWC	-	-	-	3641.39
	Hydrocarbon type	Gas	Gas	Gas	Oil and gas
Volume	STOIIP [*10^6 sbbl]	-	-	-	214
	GIIP [*10^6 sm3]	31091	23682	22881	75557
Facies	Pro delta (%)	43.9	34.8	29.8	0.26
	Channel (%)	10.89	7.65	16.49	16.59
	Mouth bar (%)	21.62	24.13	27.18	48.58
	Delta plain (%)	16.39	14.16	24.77	27.51
	Delta front (%)	7.17	19.26	1.74	7.06

(b)	Facies	Mean	Standard deviation	Minimum	Maximum
Effective model porosity	Pro delta	12.44%	2.83%	6.90%	24.35%
	Channel	19.22%	3.85%	1.14%	25.88%
	Mouth bar	17.59%	4.87%	9.67%	28.54%
	Delta plain	17.71%	4.20%	9.34%	26.13%
	Delta front	13.83%	20.30%	10.22%	19.49%

Colour code

(red)	Low	(green)	High

Away from the mouth bar and the channel is the finer delta front sediments deposited at the interphase between the proximal channel/mouth bars, and the distal pro delta which consists of predominantly argillaceous sediments. On the other hand, the channel also becomes finer away from the center towards the frequently flooded area of the basin (delta plain; Fig. 2.6). The described trends have been implemented in the model by using respective geometric parameters in the frame of the object modeling workstep.

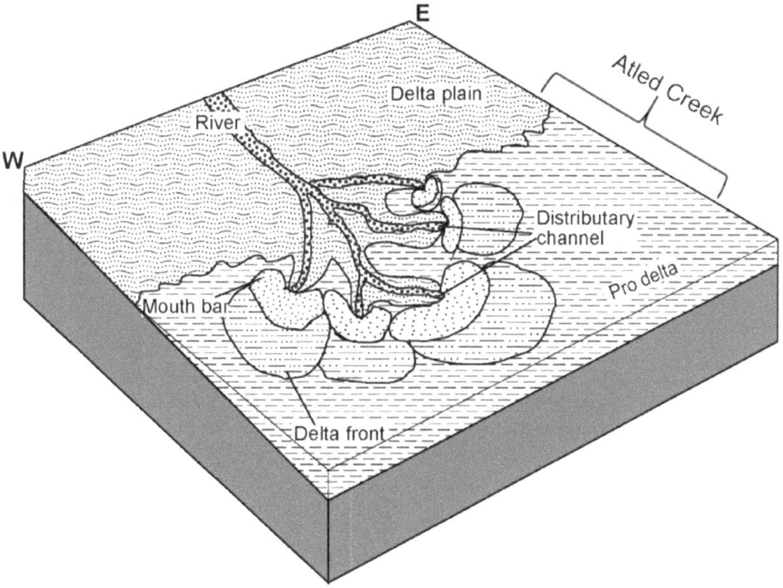

Figure 2.6. Conceptual depositional model for delta systems, showing the identified depositional facies/environments.

2.5.3 Structural elements and the structural model

Normal faults which are sub-parallel to the present-day coastline (NW – SE) are present in most parts of the field (Fig. 2.7, 2.8 supplementary materials 2.2, 2.3). A total of twenty-six normal faults have been interpreted and modelled. The seismic interpretation and the fault model (Fig. 2.7 – 2.9; supplementary materials 2.2, 2.3) reveal predominantly W–E trending faults that dip mainly in a southern direction (Fig. 2.8). A prominent major fault (F1) trending approximately WNW–ESE separates the field into two major blocks (north and south) with the existing wells located in the southern block (hanging wall). This fault curves and changes from the W-E trend in the western part to the NW-SE trend in the southern part of the field (Fig. 2.10 b).

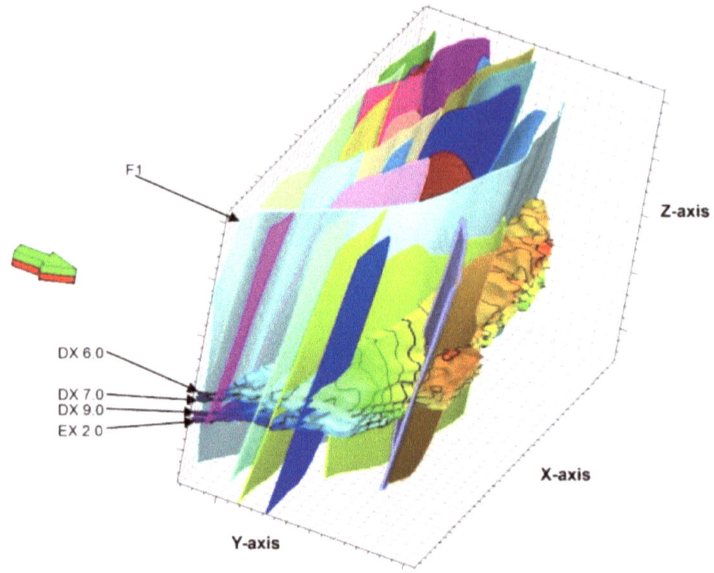

Figure 2.7. The fault model displayed on the generated structural maps (DX 6.0, DX 7.0, DX 9.0, and EX 2.0) in 3-D shows a dominantly W–E trend of the faults. Z-axis scale in meters.

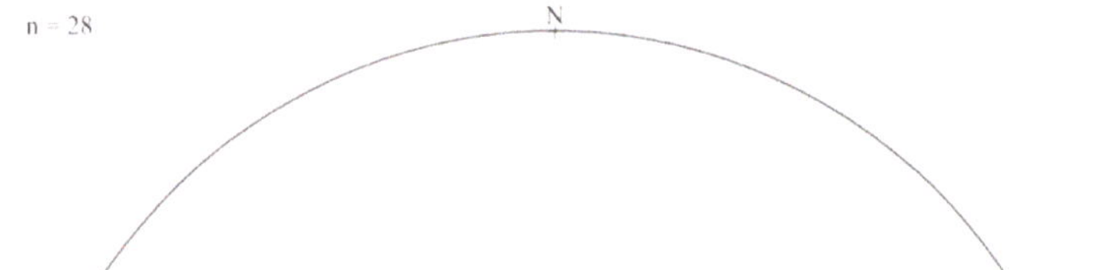

Figure 2.8. Rose plot of the interpreted faults showing the predominant WNW – ESE orientation.

The depth to the top of reservoirs (and the corresponding TWT) ranges from 2700.0 – 3701.7 m (2333.58 – 2999.31 ms), 2764.75 – 3764.13 m (2384.54 to 3043.66 ms), 2841.33 – 3963.55 m (2413.41 to 3132.13 ms), and 2888.74 – 4051.85 m (2451.83 to 3158.48 ms) respectively for DX 6.0, DX 7.0, DX 9.0, and EX 2.0.

The structural maps show the structural elements of the mapped reservoir units and indicate that the individual horizons contain an anticline roughly at the center of the concession (Fig. 2.9 a-d). Structural highs (low TWTs) are present in the northern and central parts of the field while structural lows (maximum TWTs) are situated in the western, southern, and south-eastern regions. The highest area is present north of the major fault F1.

The 3-D structural model of the field shows a fairly complex structural setting with nearly consistent thicknesses of the reservoir units (Fig. 2.10). The fine division of the reservoir units into 25 (DX 6.0), 41 (DX 7.0), 14 (DX 9.0), and 14 (EX 2.0) layers and the created 3-D grid (100 m by 100 m in each cell) enable the stochastic distribution of properties and ensure the representation of properties (value) in each cell of the grid.

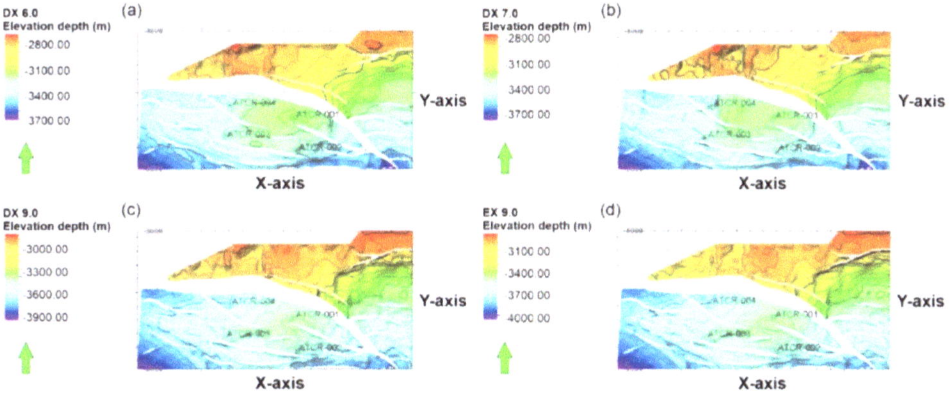

Figure 2.9. a – d) Depth-structure maps of the Miocene sandstone reservoir horizon/units (DX 6.0, DX 7.0, DX 9.0 and EX 2.0 respectively) within the "Atled Creek" concession showing structural features at the top of each reservoir.

This structural model shows the basic geometry of the concession with a notable anticlinal structure related to the normal fault (F1) in the central part of the studied area around the wells (Figs. 2.9, 2.10). Three regional fault blocks are recognized in the field based on the connectivity of the faults in the model (Fig. 2.10 b). All the existing wells are situated within the region 1 block in the southern part of the field. The northernmost part of the region 2 fault block is limited by the lack of adequate data coverage. The structural state of this region cannot be fully ascertained despite showing the possibility of anticlinal structure in the footwall of the F1 fault (Figs. 2.9, 2.10).

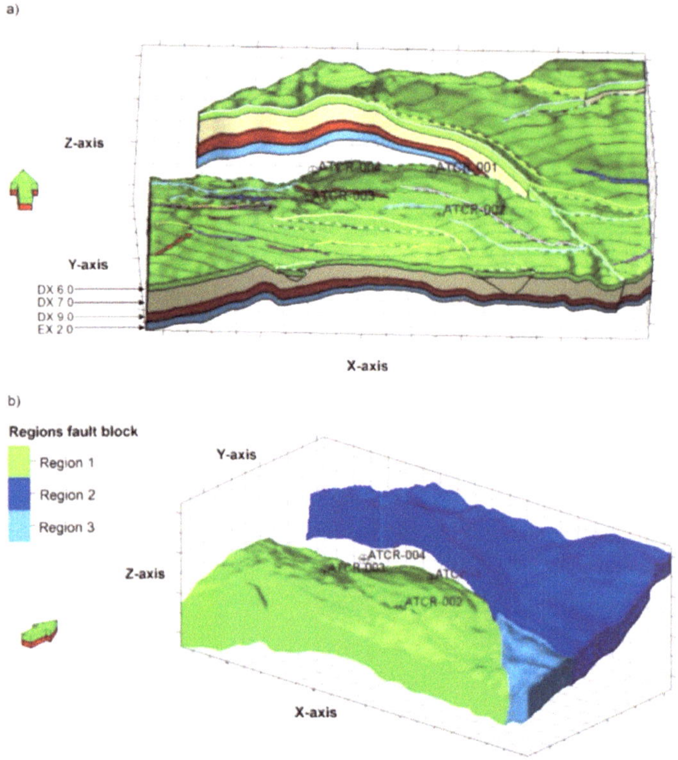

Figure 2.10. a) The structural model of the stacked reservoirs; DX 6.0; DX 7.0, DX 9.0 and EX 2.0. constructed from the integration of fault sticks and horizons. b) 3-D structural model showing the fault blocks in the field; with the existing wells located in the southern block (region 1).

2.5.4 Facies distribution and the facies model

The lithostratigraphy of the field is characterized by alternating sandstone, siltstone, and claystone sequences (Fig. 2.5). The studied wells (ATCR-001, ATCR-002, ATCR-003, and ATCR-004) penetrated both the Benin (continental) and Agbada Formations (paralic). The Benin Formation is predominantly composed of sandstone with intercalations of thin shales and does not constitute part of the further analysis and interpretation in this research as no hydrocarbon plays are associated with the formation. The base of the Benin Formation is interpreted at 2008.78 m, representing the first gamma-ray peak and high resistivity which indicates an entrance into the paralic Agbada Formation from the predominantly fluviatile Benin Formation.

The Agbada Formation is made up of alternating sandstone, siltstone, and shale sequences of varying thicknesses (Fig. 2.5). These are indicated by variably low and high gamma ray signals. The deepest point probed by the studied wells is 4,000 m while the logs are available up to 3,900 m within the Agbada Formation. The shaliness of this formation is higher in the southern part of the field (e.g. in well ATCR-002).

The depositional facies within the identified reservoirs in the central swamp subbasin of the Niger Delta vary between distributary channel, mouth bar, delta plain, delta front, and pro delta (Fig. 2.5). Channel deposits are represented by a generally fining-upward sequence deposited in the proximal region of the basin (Fig. 2.5). Mouth bar deposits are attached to the distributary channels in the proximal area and show coarsening upward signature in GR logs (GR decreases upwards). Delta plain deposits consist of finer sediments restricted to the flooded area of the channel deposits. The delta front deposits are attached to the mouth bar at the interface between the proximal and distal regions. Pro delta deposits are present in the distal region of the basin (predominantly argillaceous) and show a generally fining upward GR log signature (gamma ray increases upwards, Figs. 2.5, 2.6). Sediments deposited in the hanging wall of some faults (e.g. F1; Fig. 2.11 e, f) are thicker than those in the footwall of the fault, indicating syndepositional activity of the fault systems. It should be noted that these faults most likely has impacted the facies distribution in the area. Still we neglected the impact of syntectonic subsidence differentiation on facies distribution in our model, as we concentrated our interest only in the reservoir in the downfaulted hanging wall of the structure and we were missing data on facies proportions in the footwall. The global facies proportion of pro-delta, channel, mouth bar, delta plain, and delta front sediments vary within the generated 3-D model based on the well data and by unit from 0.26 – 43.9% (avg. 27.2%), 7.65 – 16.6% (avg. 12.9%), 21.6 – 48.6% (avg. 30.4%), 14.2 – 27.5% (avg. 20.7%), and 1.7 – 19.3% (avg. 8.8%) of the whole studied volume (Table 2.1). The calculated volume of shale (Vsh) shows gradual changes in the shaliness within the reservoirs. The 3-D facies model shows lateral and vertical facies changes in the depositional setting and the corresponding lithology of the field. Each reservoir unit is characterized by a shallowing upward succession with the pro-deltaic sediment largely concentrated at the base of each reservoir unit and more towards the southern, more distal part of the field (Figs. 2.5, 2.11). Prominent sand-rich reservoir bodies (channel and mouth bar) are more common in the deeper reservoir unit (EX 2.0; Fig. 2.11), which forms the main economic reservoir in the block. Generally, the reservoir facies bodies appear to be well laterally connected.

Prior to the implementation of the geological concept of the 3-D facies model the facies log were upscaled into the 3-D grid with the most-of algorithm, ensuring that each cell with interpreted facies logs receives a facies value of the dominating facies in its interval. According to our conceptional model we set the pro-delta facies as the background in the subsequent stochastic object modelling process.

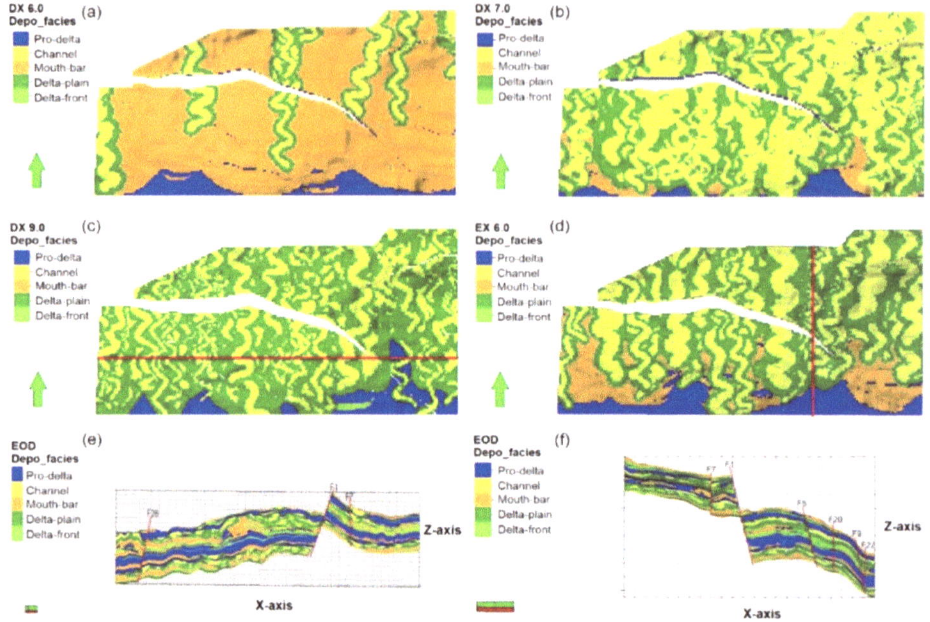

Figure 2.11. a –d) 3-D facies model of reservoirs DX 6.0 DX 7.0, DX 9.0, EX 2.0. e, f) West – east (red line shown in c) and north-south (red line shown in d) cross-section of the stacked reservoir model showing the modelled lateral and vertical facies distribution.

Using the add new channel in the object modelling pane, we added the channel facies (channel sand) and the delta plain was attached as a levee. Adjusting the geometry of the channels it was possible to match the relative amount of sandy channel and levee facies observed in the wells. Additional geometric bodies corresponding to the natural geometry of the mouth bar and delta front (as determined in the conceptual model; Fig. 2.6) were added to modelled facies, and the rule was set to replace other facies. We attached the mouth-bar and delta-front bodies to channel objects, by defining source points were channel bodies end and mouth-bar bodies are seeded.

The fraction of each facies was derived from the upscaled facies log and a triangular distribution was used to specify the minimum (min.), mean, and maximum (max.) facies proportion probability for each reservoir interval. Reasonable minimum and maximum facies proportions are obtained from the minimum and maximum fractions occurring in all individual studied reservoir intervals.

The vertical function obtained from the data analysis was used to constrain the trends in each facies distribution. The model was set to output the directional, depth trend, object distance and object curvature (Fig. 2.12) which are used to constrain the distribution of properties in the petrophysical models. We generally can assume that coarser sediments with better petrophysical properties are deposited in the area of higher curvature of channels and in the deeper parts of the channels (where velocity is reduced). Deposition of finer and hence less porous, more shaly sediments progressively takes place away from the center of the channels into the delta plains. The mouth bars have coarser, better reservoir sands deposited close to the channels and become finer seaward into the delta front. Pro-deltaic sediments are deposited at the base of each reservoir unit and represent the finest sediments with relative poorest porosity and permeability.

2.5.5 Petrophysical model

Detailed petrophysical parameters of each studied reservoir unit are presented in Table 1. Total porosity in these reservoir units range from 10.9 – 24.7% (DX 6.0, avg. 16.6%), 6.3 – 33.1% (DX 7.0, avg. 22.7%), 3.1 – 42.8% (DX 9.0, avg. 19.7%), and 6.4 – 41.7% (EX 2.0, avg. 21.8%) while the effective porosity ranges from 3.3 to 24.7% (DX 6.0, avg. 12.7%), 9.2 – 29.5% (DX 7.0, avg. 16.7%), 4.2 – 38.2% (DX 9.0, 13.3%), and 9.0 – 28.0% (EX 2.0, 18.2%). Permeability (k) ranges from 23.3 – 271.1 mD (DX 6.0, avg. 99.7 mD), 14.4 – 464.6 mD (DX 7.0, avg. 105.3 mD), 1.3 – 1676.5 mD (DX 9.0, avg. 99.4 mD), and 13.3 – 396.4 mD (EX 2.0, avg. 126.7 mD). The water saturation (Sw) ranges from 20 – 100% (DX 6.0, avg. 37.9%), 20 – 100% (DX 7.0, avg. 78.4%), 20 – 100% (DX 9.0, avg. 58.8%), and 2 – 100% (EX 2.0, avg. 36.3%) while the corresponding hydrocarbon saturation ranges from 0 to 80% (DX 6.0, avg. 62.1%), <0.1 – 80% (DX 7.0, avg. 21.6%), <0.1 – 80% (DX 9.0, avg. 41.2%), and <0.1 – 98.0% (EX 2.0, avg. 63.7%).

The petrophysical logs (porosity, permeability, water saturation) were upscaled into the 3-D grid and modelled using the Gaussian random function simulation, with co-kriging. Co-kriging allows the use of a secondary variable (volume trend) with input from previously generated depth trend, object curvature and distance to object from facies model (Fig. 2.12) to constrain the property distribution of petrophysical variables (see chapter 2.5.4). In our model we assume that there is a fair to good relationship between the facies trends and petrophysical properties and used a generic correlation factor of 0.8.

Our 3-D petrophysical models (porosity, permeability, and water saturation) show the lateral and spatial variation in the quality of the identified reservoirs (Fig. 2.13, 2.14, 2.15). As previously described the high-quality sand is modelled at the point of maximum curvature of the channels (coarse sediment, better porosity). Highly porous and permeable zones exist within the four reservoir units (Fig. 2.13, 2.14).

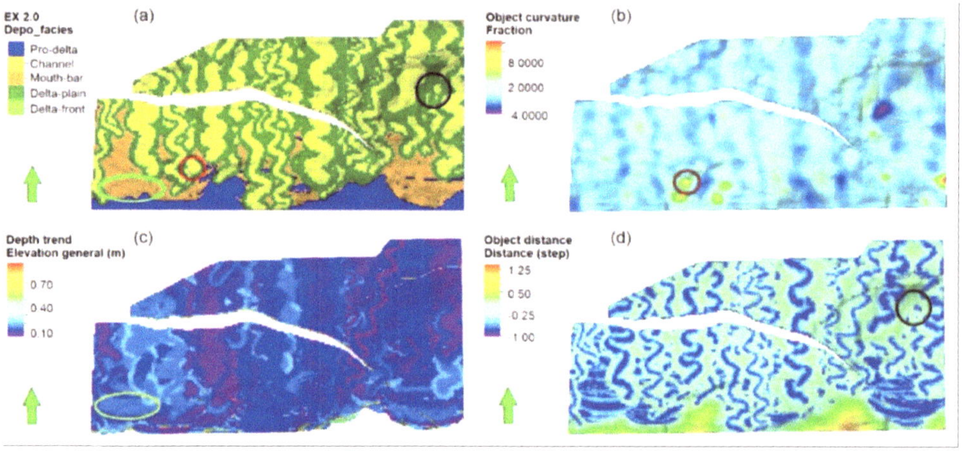

Figure 2.12. a) 3-D facies model of the EX 2.0 reservoir unit. b – d) 3-D models of the volume trends generated from the facies model (a). b) Object curvature model, with a high curvature area indicated by the red circle. c) Depth trend showing the depth variation of the property (e.g. the channel is mostly shallower than the mouth bar; green ellipse). d) Object distance showing variation in property away from the center of the channel. These (b – d) are used to constraint the petrophysical models for better property distribution.

The stacked model gave effective porosity values ranging from 6.9 – 24.4% (mean 12.4%), 1.1 – 25.9% (mean 19.2%), 9.7 – 28.5% (mean 17.6%), 9.3 – 26.1% (mean 17.7%), and 10.2 – 19.5% (mean 13.8%) in pro delta, channel, mouth bar, delta plain, and delta front respectively (Table1). However, reservoir units DX 6.0 and DX 9.0 has the lowest effective and total porosities, and permeability also reaches its lowest in these reservoir units despite occasionally highly permeable zones (Fig. 2.13, 2.14). The water saturation (from the model) reached its highest (100%) at the flanks of the anticlinal structure located at the center of the field, while it remains low in the center of the anticlinal region (Fig. 4.14). The petrophysical analysis and 3-D model of water saturation show lower values in reservoir EX 2.0 (Fig. 2.15, Table 2.1).

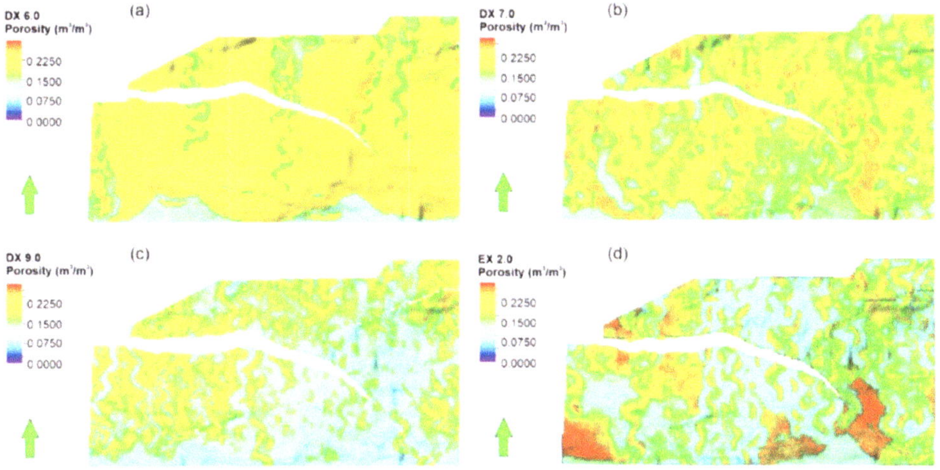

Figure 2.13. Porosity model of the Miocene reservoirs DX 6. 0 (a), DX 7.0 (b), DX 9.0 (c), and EX 2.0 (d). The facies-controlled distribution of porosity can be seen as fluvial channels and mouth bar deposits have higher porosities than pro-delta and delta plain deposits (cp. Fig. 2.9).

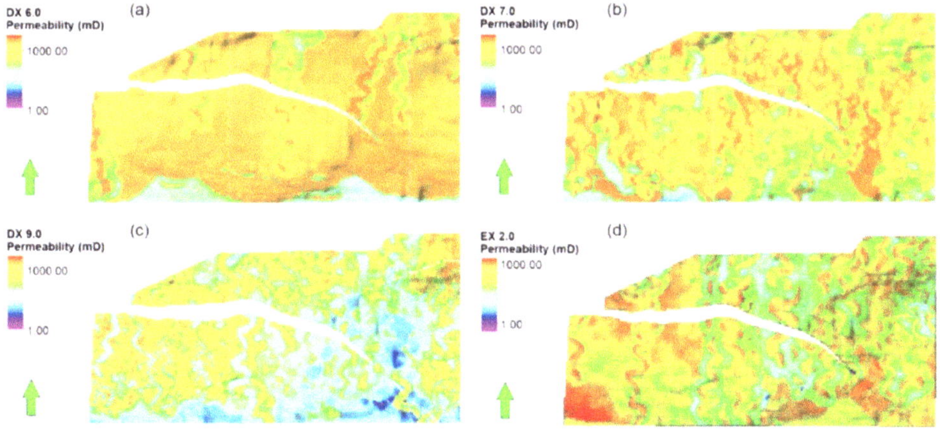

Figure 2.14. 3-D model showing the distribution of permeability in reservoirs DX 6. 0 (a), DX 7.0 (b), DX 9.0 (c) and EX 2.0 (d).

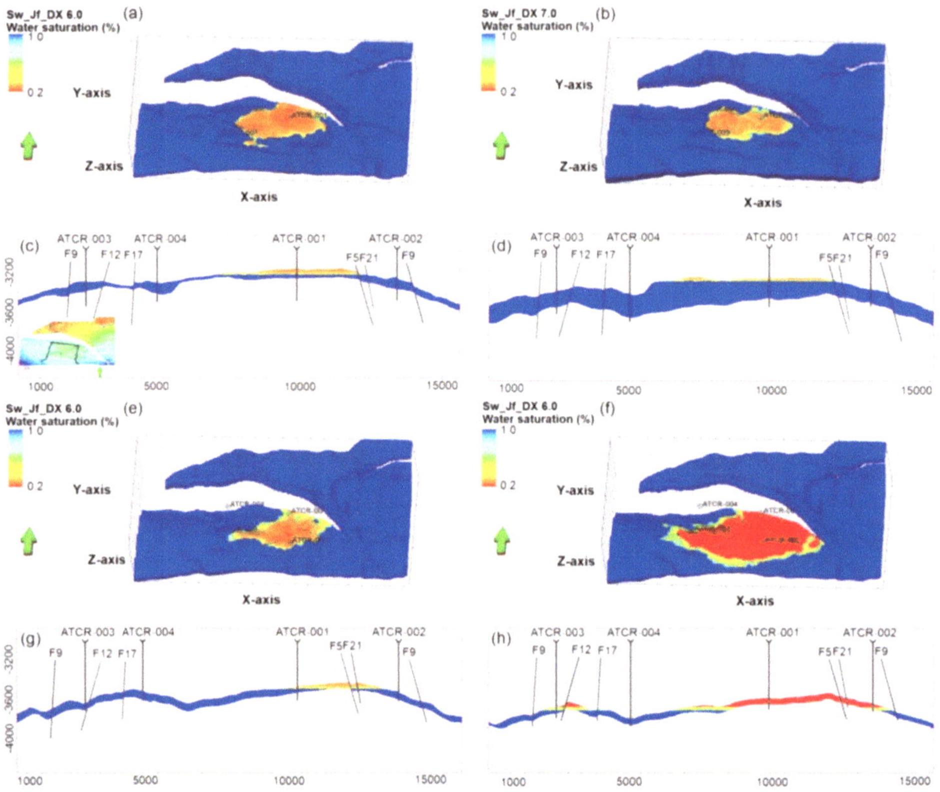

Figure 2.15. The 3-D water saturation model (a, b, e, f) and cross-sections (c, d, g, h) of the Miocene reservoirs in the Niger Delta.

2.5.6 Hydrocarbon contacts and reservoir volumetrics of the reference model

The gas water contact (GWC) in reservoir units DX 6.0, DX 7.0, and DX 9.0 are present at depths of 3232.26 m, 3294.37 m and 3484.29 m respectively while a gas-oil contact (GOC) and oil water contact (OWC) occurs at a depth of 3602.69 m and 3641.39 m in reservoir EX 2.0 (Fig. 2.16). The hydrocarbon volume in the "Atled Creek" concession is quantitatively calculated from the static model (Table 1). In the reference static model, the calculated volume of gas initially in place (GIIP) is 31,091*10^6 sm^3, 23,682*10^6 sm^3, 22,881*10^6 sm^3, and 75,557*10^6 sm^3 (where sm^3 = standard cubic meters) respectively in reservoirs DX 6.0, DX 7.0, DX 9.0, and EX 2.0. Stock tank oil initially in place (STOIIP) is calculated to be 214*10^6 sm^3 in the only oil-bearing reservoir EX 2.0 (Table 2.1).

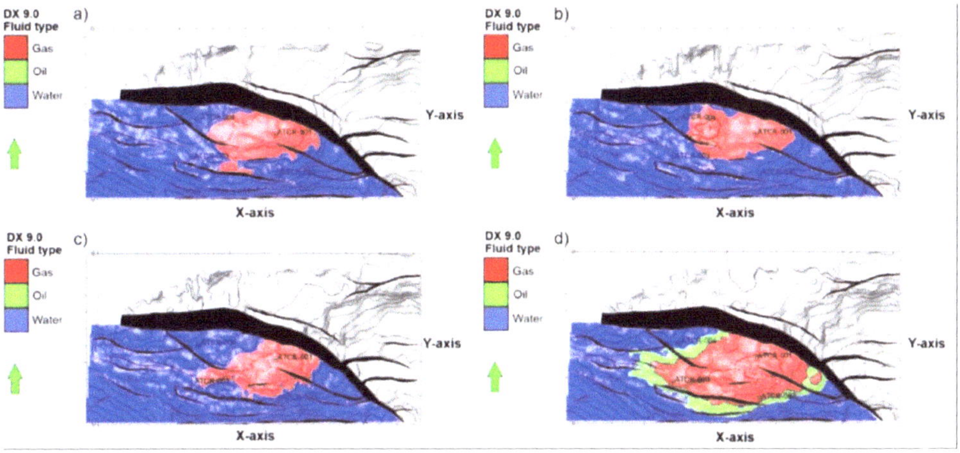

Figure 2.16. Generated fluid contact maps showing gas and water in reservoirs DX 6.0 (a), DX 7.0 (b), and DX 9.0 (c), and gas, oil and water in reservoir EX 2.0 (d).

2.5.7 Sensitivity and uncertainty

Based on the petrophysical interpretation of the individual reservoirs and available regional data (Lambert-Aikhionbare, 1981; Oboh-Ikuenobe et al., 2005; Oboh, 1995), values were assigned for petrophysical and proportion variables in the sensitivity and uncertainty quantification (Table 2.2).

In the exploration phase the main uncertainty associated with the volumetric calculation of hydrocarbon in place based on the 3-D static model are usually those that may be classified as being associated with structures and reservoir properties.

The structural uncertainty includes missing faults below the seismic resolution and positioning uncertainty related to depth conversion (velocity) due to the limited data available. Our depth maps have been generated from a carefully constructed 4-layer based velocity model (using interval velocity) and they are controlled by checkshots and well picks. Different structures, depths, and thicknesses of the reservoirs may result from velocity models generated by different methods or using different data (e.g. velocity cube). This might ultimately affect the volumetric calculations. However, provided the availability of well picks we expect to have kept the presumed differences minimal. The possibility of missing faults and/ or fault depth differences might also constitute some levels of uncertainty in the volumetrics, but more so for the production behavior of the reservoir. We have tested a method to quantify this structural uncertainty, which requires a special more complex methodology. As we concentrate on static volumes, this does not constitute part of the presented work.

Accordingly, the main studied static volume uncertainties are resulting from facies proportion and petrophysical properties (porosity, permeability), which in part depend on other factors (e.g. tortuosity factor, cementation factor, sorting; see table 2.2).

Table 2.2. Range of the geological parameters used in the uncertainty and sensitivity analysis of reservoir EX 2.0.

Variables	Arguments	
	Minimum	Maximum
Tortuosity factor	0.5	1.5
Sorting index	0.7	1
Cementation factor	1.8	2
Porosity (fraction)	0.002	0.2

In our sensitivity study we must consider that the facies proportion and the facies connectivity may vary within the field and thus affect our calculations. Pore volume estimates are affected by the previously described porosity variations in the field as a response to lateral and vertical facies changes or lithology and in addition by a global uncertainty of the field wide average porosity within facies classes. Permeability was calculated from porosity with a regression model, which has some uncertainty and has a similar resulting effect on reservoir volumetrics by influencing water saturation. Similarly, other factors with impact on the saturation may vary within a narrow known limit within the basin and also investigated in the sensitivity study (Onengiyeofori et al., 2019). The free water level (FWL) of the reservoir is uncertain and may be too optimistic or pessimistic and its position was also included in the uncertainty evaluation.

From the 250 realizations based on Monte-Carlo simulation for reservoir EX 2.0, the uncertainty quantification results in the obtained low (P10), high case (P90) and mid-case (P50) volumes (Table 2.3; Fig. 2.17). The result shows variation in the stock tank oil initially in place (STOIIP) and gas initially in place (GIIP) in the model in which the obtained most-likely case is 406*0^6 bbl and 302,740*10^6 m^3 for STOIIP and GIIP, respectively. The low- and high-cases for the STOIIP are 388.1*10^6 bbl and 425.4*10^6 bbl while the GIIP has 294110*10^6 m^3 and 311169*10^6 m^3 as the low- and high-cases, respectively.

The sensitivity analysis of the EX 2.0 reservoir model shows that the generated 3-D static model is most sensitive to the tortuosity factor (TF), sorting (sort), facies (channel and delta plain, and delta front), cementation exponent (m), and porosity (Fig. 2.18). These show the relative volume changes based on each uncertain parameter in the model.

Table 2.3. Summary of the result of the uncertainty analysis in reservoir EX 2.0 with the P10; P50 and P90 ranking of the hydrocarbon volume

Ranking	P10	P50	P90
STOIIP (*10^6 bbl)	388	405	425
GIIP (*10^6 m^3)	294,110	302,740	311,169

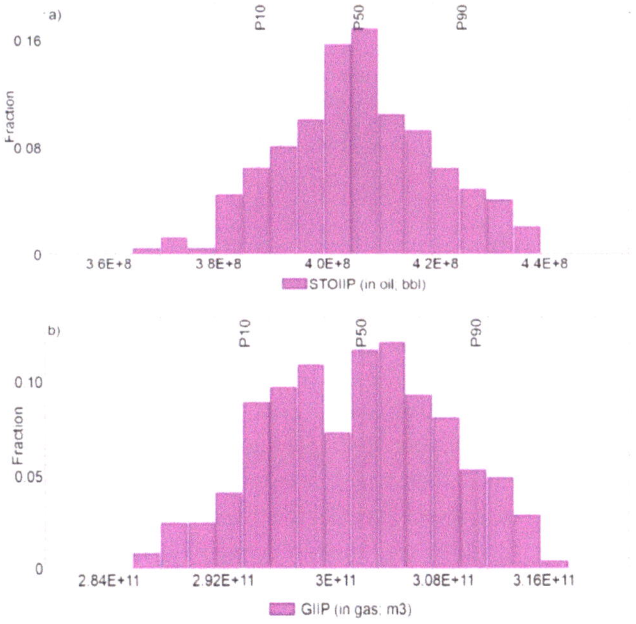

Figure 2.17. The histograms of 250 runs carried out in the frame of the uncertainty analysis of reservoir EX 2.0 with P10 (low case), P50 (most likely) and P90 (high case) for a) STOIIP and b) GIIP. STOIIP = stock tank oil initially in place, GIIP = gas initially in place.

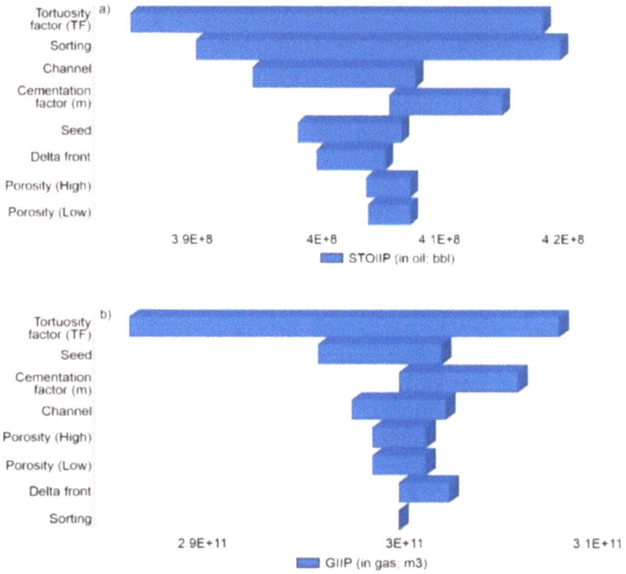

Figure 2.18. Tornado plot of the sensitivity of gas-in-place and oil-in-place of the EX 2.0 reservoir model based on uncertainty in investigated geological and petrophysical parameters. This results in a range of volume captured in the volumetrics calculation. STOIIP = stock tank oil initially in place, GIIP = gas initially in place. Note that the studied uncertainty tortuosity factor has the highest impact on reservoir volumes.

2.6 Discussion

2.6.1 Depositional facies and environment

The depositional facies in the "Atled Creek" concession varies laterally and vertically between pro delta, channel, mouth bar, delta plain and delta front. These rapid variations are reflective of sea-level fluctuation which resulted in alternating proximal (coarse) or distal sediment (fine) deposition and account for the paralic nature of this deltaic basin (Figs. 2.5, 2.6, 2.11). Consequently, the lithology of the field varied rapidly between sand, silt, and shale. Sand deposits are restricted to the channel and mouth bar facies while the pro delta is predominantly made up of shale. On the other hand, the delta plain, and delta front are mainly silty with intercalated shale (Fig. 2.5). Generally, this is consistent with the description of Oboh (1993); (Oboh, 1995) for the "E2.0" reservoir unit of the Central Swamp subbasin. The calculated volume of shale shows gradual changes in the shaliness of the reservoirs (Table 2.1). This is a reflection of the changes in grain size which are largely controlled by lithofacies and depositional environments (Nton & Salami, 2016; Oboh, 1992a, 1993; Selley, 1970). The object-based stochastic approach to facies modelling has proved to be a valuable input for the petrophysical model in which the property distribution is better controlled.

Generally, the depositional facies distribution in the model allows the interpretation of the dominant geological events that occur during deposition. The transgressive and regressive cycles are revealed by the facies model from the bottom to the topmost reservoir units (Fig. 2.11 a – d). Deposition of the EX 2.0 reservoir unit is interpreted to occur during a regression when sandy (coarse) sediments were laid down. This regressive cycle is succeeded by a transgression during which pro-deltaic sediment dominated the deposition in the DX 9.0 reservoir unit. During the deposition of the DX 7.0 reservoir unit, the transgression continued but slowly with lesser pro-deltaic sediments lain down compared to DX 9.0. The basin returned to the regressive phase during the deposition of the topmost part of the DX 6.0 reservoir unit and sandier deposits (channel and mouth bar) were lain down (Fig. 2.11 a – d).

2.6.2 Reservoir quality

We can observe that the porosity is dependent on the reservoir facies. The channel sediments show the highest porosity while the pro deltaic sediments show the lowest effective porosity in the model (Table 2.1). Average porosity per reservoir unit is believed to be largely controlled by the amount of pro-deltaic shale present as a result of relative presence of the depositional environments. Furthermore, the variation of depositional environments affects the sedimentological properties including grain size, sorting, and cementation. Detrital and authigenic kaolinite, calcite, dolomite, and siderite have been reported as the major cementing minerals in the Niger Delta (Lambert-Aikhionbare, 1981; Oboh, 1992a). The clay-rich sandstone facies often shows moderate to high porosity but low permeability in the Niger Delta and neighboring Dahomey Basin (Adamolekun et al., subm.; Adamolekun et al., 2022; Lambert-Aikhionbare, 1981). This is held responsible for the observed low permeability intervals in reservoir units DX 6.0 and DX 9.0, although porosity is comparable to the other reservoir units. We have included the reservoir quality in our study to represent diagenesis with the tortuosity and cementation factors impacting the permeability and water saturation estimates.

2.6.3 Trap types and structures

The 3-D structural model shows a fairly consistent fault pattern dominated by normal faulting in a gently subsiding basin. This is supported by the regional structural dynamics and tectonics of an extensional regime also documented by Doust and Omatsola (1990); Khani and Back (2015); Ogbe and Edegbai (2022). The thicker sequence of sediment in the hanging wall of the normal faults and the ratio of the footwall thickness to the hanging wall thickness of greater than 2 along the major fault F1 shows that the faults grow during deposition (growth faults; Fig. 2.11 e, f). This is a reflection of structurally controlled accommodation space that increases basinward. The characteristic normal faults and roll-over anticlines in the field are evidence of suitable trap formation. The anticlinal structures owe their origin to shale diapirism or are associated with the accompanying rollover structures in the Niger Delta region (Magbagbeola & Willis, 2007). The faults are largely syndepositional structures and are established during the progradation of the delta (Ibe & Ezekiel, 2018). Generally, the target of hydrocarbon exploration in the basin are mainly the roll-over anticlines, faults, and 4-ways dip closures (anticlines) (Adeigbe & Alo, 2017; Doust, 1990; Magbagbeola & Olayinka, 2019; Magbagbeola & Willis, 2007). The extensional system in this region was initiated by the separation of African and South American plates in the Lower Cretaceous (Doust & Omatsola, 1990; Reijers, 1996, 2011). The rifting diminished in the Late Cretaceous and gravity tectonism became the primary deformational process (Tuttle et al., 1999). Growth faults, with basinward dip directions, are also prominent in other deltas, e.g., the Nile and the Mississippi Delta (Abd-Allah et al., 2012; Day et al., 2020; Pigott & Abdel-Fattah, 2014). We can observe these growth faults in our seismic interpretation and the extension due to shale diapirism in the crestal graben-like structures in the roll-over structure of the hanging wall, which forms our studied trap.

2.6.4 Prospect and leads

With the help of 3-D structural and petrophysical models, unexplored areas of a basin and/or field can be properly evaluated. This provides valuable background for future exploration, expansion, optimization, and development. The structural geometry shows anticlinal traps with structural closures at the centre of the concession. The concession contains 28 identified faults, which could result in the compartmentalization of the reservoirs (Figs. 2.7 – 2.10; supplementary material 2.2). Deeper reservoir units (EX 2.0) show prospects for future development and reservoir optimization. Provided that the main structural crest has been first produced, prospective areas for future development fall roughly in the southern part of the current well locations (Fig. 2.19, black box). In this study, the static 3-D model reveals potential compartments, which could accommodate further production wells through the evaluation of the geometrical distribution of reservoir properties and structures. Production in the "Atled Creek" concession is expected to be enhanced if additional wells are drilled within the region's fault block 1. The overall reliability of the trap is proven by the observed high hydrocarbon saturation zone lying above the mapped spill point (leakage) in the water saturation model (Figs. 2.15, 2.16). The northern block of the field (beyond the F1 regional fault) remains a lead as the region lacks well control and adequate data. While there seems to be a high possibility of closure(s) in the area, the fault system could not be fully explored beyond the available data set and thus, we classify the area as a lead. Additional seismic and well log data and further inclusion of charge history would aid the identification of other prospects in the region.

The generated 3-D grid data could serve as a primary data source in reservoir and production simulation and management. Production simulation studies can be conducted accurately based on the generated 3-D structural and property model.

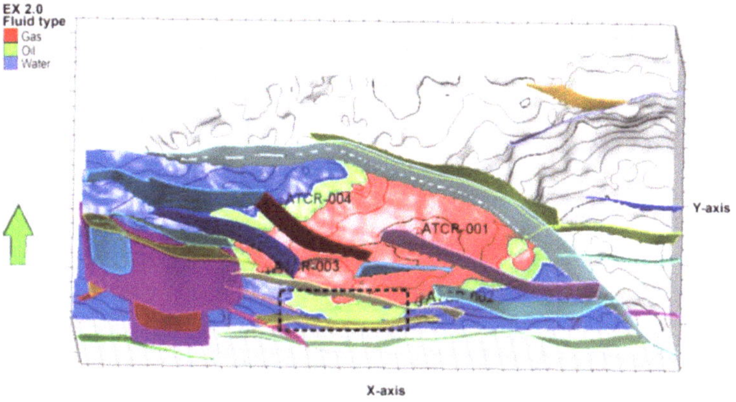

Figure 2.19. The fault model displayed on the fluid contact map shows a promising area for future development (dashed black box).

2.6.5 Sensitivity and uncertainty

The most sensitive parameters in the generated 3-D static model are the tortuosity factor, sorting, facies, and the cementation factor (Fig. 2.18). Furthermore, the model is sensitive to global porosity variation (Fig. 2.17, 2.18). We note that the tortuosity factor, a dimensionless property relating the flow path to the flow distance (Ghanbarian et al., 2013; Kou et al., 2012) is affected by diagenetic modifications, which could either decrease or increase the actual flow path at pore level in the reservoir (e.g. due to cementation or dissolution). A high tortuosity factor (≥ 1.0) caused lower permeability and higher water saturation which yielded lower volume of hydrocarbon in the model. Our model also shows that realizations with better sorting index (≤ 0.7) results in higher volumes of hydrocarbons. Sorting is the measure of variability in the grain size of sedimentary rocks (i.e. the measure of spread around the mean) (Boggs Jr., 2009; Folk & Ward, 1957). It is observed that the relative amount of facies in the model influences the volume of hydrocarbons in the model. High amounts of channel and mouth bar facies tend to favor high volume of hydrocarbon, related mostly to the mean better porosity of these facies associations (Table 2.1). However, provided the current uncertainty, the impact on volumetrics of porosity is low compared to the influence of the tortuosity factor, sorting, facies, and cementation factor.

Hence, incorporating data from core analysis to constrain these factors (tortuosity, sorting) would help in reducing the uncertainty in the petrophysical analysis and in turn yield a more reliable quantification of our reservoir fluids. Also, the model is not fully adequate for the areas with no data (regions fault blocks 2 and 3; Fig.2.10 b). Thus, the availability of more wells or regional facies distribution maps will help to overcome the uncertainty associated with facies changes that might have affected our calculations. Furthermore, the availability of stacking velocity or a velocity cube would help to overcome the structural uncertainty that might have been related to depth conversion.

2.7 Conclusions

The series of transgression and regression cycles account for the deposition of different packages of proximal and distal sediments in the basin. The highest effective porosity is obtained in the distributary channels while the pro delta is at the lowest value. Prodeltaic and delta front sediments are the most important local and regional seals in the basin while the channel and mouth bar deposits contain high reservoir quality sediment. Facies models generated by the stochastic object-based approach are important in revealing the heterogeneity of the reservoirs. Vertical proportion curves generated from data analysis better constrain the distribution and the connection of each facies in the model using a stochastic (object-based) approach. The facies model indicates the reservoir units are dominated by sandstone and siltstone with subordinate shale in this deltaic setting. The Gaussian random function simulation method, with the co-kriging option, is effective in capturing the variations in the petrophysical properties of the reservoir by using the volume trends as secondary inputs. Normal faulting systems and the accompanying roll-over anticlines typical of extensional regimes, constitute a major structural feature of this region. The SW area of the reservoir shows more deformation than the other parts of the field. Judging from the structural framework, good trapping mechanisms are interpreted to be present within the field. The effect of the depositional facies on petrophysical properties is accurately captured using the stochastic object-based modelling approach.

Generally, the available data reveal the presence of good reservoir rocks in the studied field. Comparing the petrophysical, facies, and structural models, we conclude that unit EX 2.0 constitute the best reservoir quality in the concession. The EX 2.0 reservoir unit contains oil and gas while the remaining reservoirs lying above are only containing gas. The structural and property modelling could serve as valuable input in the initiation of development strategy and subsequent simulation as it unravels variability in facies, fluid, and petrophysical properties of reservoir rocks. The uncertainty in volume estimation arises from the sensitivity of the model to the tortuosity factor, facies, cementation factor and porosity. These uncertainties can be minimized with the incorporation of core-based data and diagenetic studies into the model.

3 Petrography and reservoir quality controls in shallow transitional marine Cretaceous-Paleogene deposits in the Dahomey Basin, Nigeria

3.1 Abstract

Upper Mesozoic and Early Cenozoic sedimentary rocks are of great significance to the petroleum geology of the Eastern Dahomey Basin as they are valuable prospects for hydrocarbon reservoirs. Diagenesis and reservoir quality (porosity and permeability) controls are deduced from core samples recovered from six shallow exploration wells in the onshore area of the Eastern Dahomey Basin using petrographic, petrophysical and mineralogical tools. The studied lithostratigraphic intervals consist of limestone, calcareous siliciclastic sandstone (Ewekoro and Araromi Formations), quartz arenite and clay-matrix supported sandstone (Afowo Formation) which were deposited within a shallow marine depositional environment.

Siliciclastic samples of the Ewekoro Formation consist predominantly of calcareous sandstone and limestone, which are classified as packstone and wackestone. The porosity values in the sandstones range from 5 – 36 % whereas permeability ranges from 0.006 – 109 mD. Ewekoro Limestone porosity ranges from 7 – 32% and permeability ranges from 0.003 – 13 mD. Samples of the Araromi Formation exhibit a porosity of 4 – 45% with permeability between <0.001 and 1764 mD in sandstones whereas a single limestone sample has 8% porosity and permeability of 0.03 mD. Porosity in quartz arenitic sandstones of the Afowo Formation ranges from 15 – 47% with permeability of 0.05 to >10,000 mD, while in the clay-matrix supported sandstones, the values range from 18 – 42% and 0.05 – 5 mD for porosity and permeability, respectively. The main pore-occluding authigenic minerals are carbonate cements (Fe-dolomite, Fe-calcite, siderite and dolomite), iron oxide, kaolinite, and illite-smectite mixed layer clays. The best reservoir quality is observed in the quartz arenitic sandstones of the Araromi and Afowo Formations, except for the low permeability, clay-matrix supported sandstone facies.

The most significant controlling factors in sandstone are clay mineral matrix contents and the presence of carbonate cements. Fe-calcite, Fe-dolomite and kaolinite formation occurs during early diagenesis in the studied lithologies and has helped to reduce the effect of compaction. The results of this study indicate fair to good reservoir quality in the shallow marine sedimentary rocks of the Eastern Dahomey Basin, which could be targeted in further exploration campaigns in the region.

3.2 Introduction

Shallow marine depositional systems contain significant hydrocarbon reservoirs worldwide. Their properties respond to several driving mechanisms such as the physical regime, internal dynamics of the coast and shelf, sea level changes, sedimentation rate, geodynamics, provenance, tectonic settings, and climatic condition (Ainsworth et al., 2008; Boyd et al., 1992; Harris et al., 2002; Siddiqui et al., 2017; Walker, 1992). Reservoir quality analysis including diagenetic, petrophysical and sedimentological properties are of important significance in targeting reservoirs for oil and gas, CO_2 storage, and aquifers in geothermal projects (Becker et al., 2017; Busch et al., 2019; Monsees et al., 2020b; Morad et al., 2010; Schmidt et al., 2018; Selley, 1998).

The studied Cretaceous-Paleogene lithologies may be analogues for the reservoir lithologies in offshore Dahomey and Niger Delta Basins. Lambert-Aikhionbare (1981) noted that the modern Niger Delta extends offshore almost to the limit of the continental shelf and continues westward into the Dahomey Basin without any major geological break. The Cretaceous successions in the Niger Delta are mostly composed of shales (Egbogah & Lambert-Aikhionbare, 1980; Lambert-Aikhionbare, 1981; Short & Stäuble, 1967; Weber & Daukoru, 1975) and are rarely penetrated during drilling (Lambert-Aikhionbare, 1982). Time equivalent onshore deposits offer an opportunity to study the influence of diagenesis on the reservoir properties of the sedimentary rocks in the Eastern Dahomey Basin.

The Eastern Dahomey Basin, a frontier basin in Nigeria (also known as Benin Basin, or Dahomey Embayment) has its sedimentary fill extending from onshore to offshore, through the coastal areas of southern Nigeria with depositional environments ranging from fluvio-lacustrine over shallow marine to marine. Despite several bitumen seepages, tar sand and some oil and/or gas shows, the Eastern Dahomey Basin presently hosts only one successful production site in the Aje-Field, which started production in 2016 (OffshoreEnergyNewsletter, 2019). The basin is the focus of both academic and industrial research. However, the reservoir quality controls are not identified.

In this study, petrophysical data (He-porosity and air-permeability), petrographic and diagenetic studies in concert with XRD analyses are used to assess the controls and the evolution of reservoir quality of the shallow-marine siliciclastic and carbonate rocks in the onshore part of the Dahomey Basin.

3.3 Geological setting

The Dahomey Basin is a marginal pull-apart basin, which was developed as a result of rifting between the South American and African Continents during the Late Mesozoic following the break-up of Gondwana (McLoughlin, 2001). The sedimentary fill of the basin ranges from Cretaceous to Recent (Fig. 3.1, Fig. 3.2, Fig. 3.3). Field and subsurface evidence shows an east-west trend of each stratigraphic unit (Obaje, 2009). Details about the tectonic evolution of this basin were documented by several authors (Billman, 1976, 1992; Hoffman et al., 1974; Petters, 1978), and later reviewed by Omatsola and Adegoke (1981). Maximum sediment thicknesses are recorded at the border between Nigeria and the Republic of Benin and are bound on its western margin by faults (Billman, 1976; Omatsola & Adegoke, 1981). The eastern limit of the Dahomey Basin is defined by the Chain Fracture Zone (Fig. 3.1 inset; Omatsola and Adegoke, 1981; Whiteman, 1982) separating the Dahomey Basin from the western limit of the Niger Delta Basin (Omatsola & Adegoke, 1981). This fracture zone consists of NW-SE trending, ocean-ward-dipping extensional faults that are offset by NE-SW oriented transfer faults (De Matos, 2000), with the fault system extending from the Mid-Atlantic ridge which formed during the Aptian as a result of seafloor spreading (Davies et al., 2005). The Dahomey Basin extends from east of the Volta Delta, Ghana in the west through Togo and Benin Republic to Western Nigeria in the east and forms multiple horsts and grabens, especially in the western flank (Omatsola & Adegoke, 1981).

3.3.1 Stratigraphy of eastern Dahomey Basin

The general stratigraphy of the Eastern Dahomey Basin (Fig. 3.3) is well known from several studies but with some controversies regarding nomenclature and/or age (Nton, 2001; Obaje, 2009). The controversies are mostly due to the lack of good borehole coverage and adequate outcrops needed for detailed stratigraphic analysis (Olabode, 2006). This has led to different stratigraphic names given to the same formations at different localities within the basin (Coker, 2002). Notable studies and schemes for the stratigraphy of the Dahomey Basin were published by various authors (Adegoke, 1969; Ako et al., 1981; Billman, 1976; Coker et al., 1983; Ehinola & Oluwajana, 2016; Ekweozor, 1990; Falufosi & Osinowo, 2021; Fayose, 1970; Jones & Hockey, 1964; Nton, 2001; Okosun, 1990; Omatsola & Adegoke, 1981; Reyment, 1965).

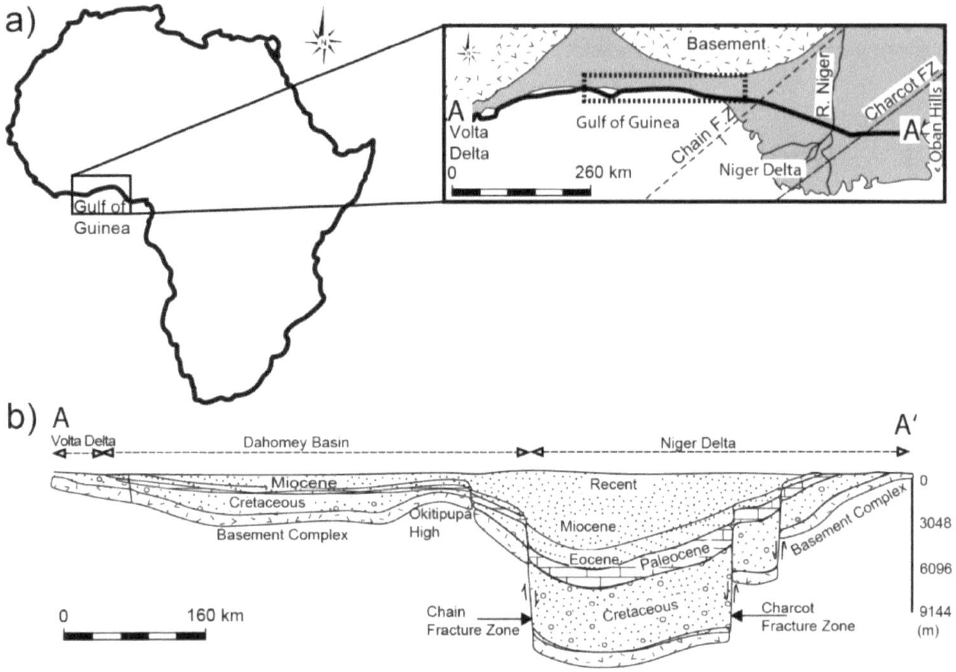

Figure 3.1. Map of Africa showing location of the Dahomey Basin. The rectangle shows the location of the study area. b) Schematic west-east cross section showing the Dahomey Basin and Niger Delta (modified from Whiteman, 1982).

The Cretaceous stratigraphy of the Eastern Dahomey Basin (Fig. 3.3) proposed by Omatsola and Adegoke (1981), which is accepted or modified by several authors is adopted in this study. Based on this, the Early Cretaceous Ise Formation lies unconformably on the Precambrian Basement Complex of Southwestern Nigeria and forms the oldest stratigraphic unit within the basin. It consists of conglomerates and pebbly sandstones (locally called grits) at the base which is overlain by coarse to medium grained loose to weakly consolidated sandstone with inter-bedded kaolinitic clay and claystone. It is interpreted as a pre-drift sequence of fluvio-lacustrine deposits ranging in age from the locally used Neocomian (probably Valanginian/Barremian) to Albian (Omatsola & Adegoke, 1981).

The Late Cretaceous Afowo Formation, which overlies the continental deposits of the Ise Formation is composed of medium-grained sandstones with inter-bedded shales, siltstones and claystones. The proportion of organic-rich shale increases progressively from bottom to top (Enu, 1990; Omatsola & Adegoke, 1981). The sandy facies are commonly tar bearing or bitumen impregnated. The lower part of this formation is transitional with mixed brackish to marginal marine intervals that alternate with well sorted, sub-rounded sands and sandstones (Elueze & Nton, 2004). They are interpreted to

represent a littoral or estuarine near-shore environment with fluctuating water level. Billman (1976, 1992) assigned a Turonian age to this formation with the upper part ranging into the Maastrichtian.

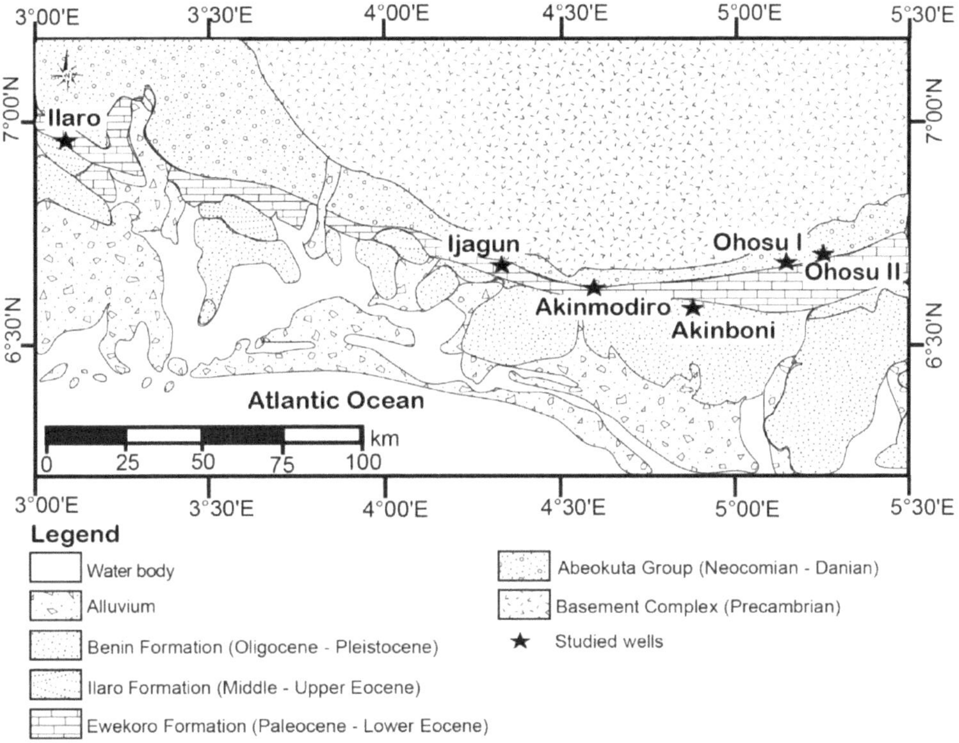

Figure 3.2. Geological map of the onshore Eastern Dahomey Basin with the locations of the studied wells (modified from Billman, 1976).

The youngest Cretaceous sedimentary rocks in the Eastern Dahomey Basin belong to the Araromi Formation, which is composed of fine- to medium-grained sandstones at the base, overlain by shales and siltstones with interbedded limestones, marls and lignites. The shales are light grey to black, organic-rich, mostly marine and range in age from Maastrichtian to Paleocene (probably Danian; Omatsola and Adegoke, 1981).

Above the youngest Mesozoic unit is the Paleocene to Eocene Ewekoro Formation which is predominantly composed of limestones. At the base, it is sandy and grades laterally into the underlying Araromi Formation (Nton, 2001). The limestones are laterally extensive and are traceable over a distance of about 320 km. The limestones are shallow-marine deposits with an abundance of marine fossils: coralline algae, gastropods, pelecypods, echinoid fragments and other skeletal debris (Fayose, 1970; Nton, 2001; Whiteman, 1982a).

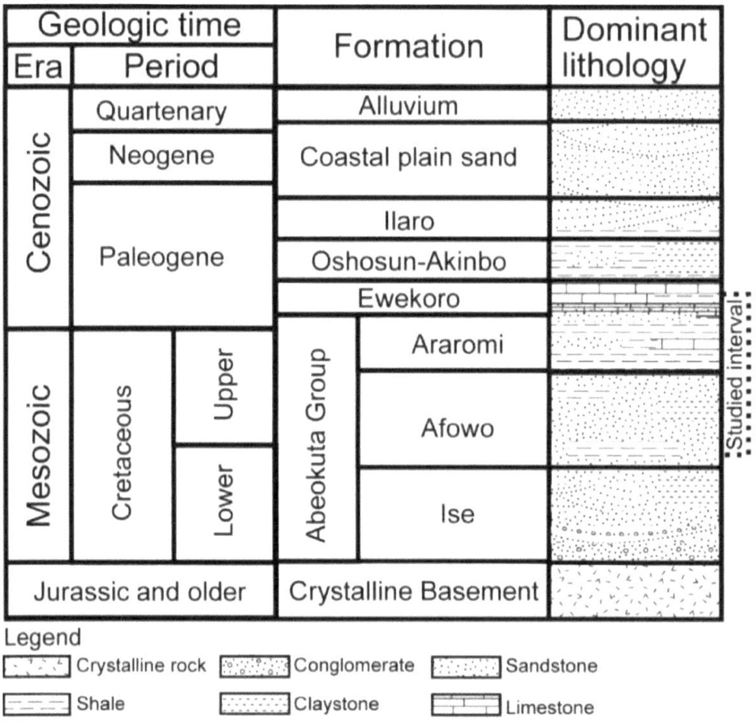

Figure 3.3. Simplified stratigraphic units of the Eastern Dahomey Basin showing the lithostratigraphic units (Billman, 1976; Ako, 1980; Omatsola and Adegoke, 1981; Agagu, 1985). Lithological symbols follow the USGS standard.

The Paleocene to Eocene Akinbo Formation (Ogbe, 1972) consists of a shale and claystone sequence. The base of the formation is defined by the presence of a glauconitic bed (Nton, 2001). East of Ijebu Ode, the formation replaces the Ewekoro Formation, which thins out there. However, it extends westward into the Republic of Benin and Togo (Slanky, 1962). In the field, the shales are grey, fissile, clayey and concretionary, dipping gently (<5°SW; Nton, 2001).

The Oshosun Formation consists of greenish-grey or beige clay and shale with interbeds of sandstones. The shale is thickly laminated and glauconitic. According to (Okosun, 1998), the basal beds consist of sandstones and mudstones, which are mostly phosphate-bearing (Nton, 2001). Additionally, some intervals can be classified as phosphorites (Nton, 2001), which were inferred to have been deposited during an Early to Middle Eocene marine transgression (Ypresian – Lutetian; Okosun, 1998).

The overlying Ilaro Formation consists of thick, yellowish, poorly consolidated, cross-bedded sandstones and kaolinitic clay which are of Middle to Upper Eocene (Nton, 2001).

The Benin Formation/Coastal Plain Sands overly the Ilaro Formation and consists of poorly sorted, partly cross-bedded sands with lenses of clays. It shows characteristic features of transitional to continental depositional environments (Agagu, 1985; Ako et al., 1981; Billman, 1976; Nton et al., 2009; Omatsola & Adegoke, 1981). This formation is dated to the Miocene to Recent (Whiteman, 1982a).

Alluvium deposits consisting of unconsolidated sand, silt and clay overlay the sedimentary units in Eastern the Dahomey Basin (Agagu, 1985; Falufosi & Osinowo, 2021). The study of calcareous nannofossils yielded a Pleistocene age in a shallow marine setting (Adamolekun & Umoh, 2016).

3.4 Materials and Methods

Core samples were recovered from six shallow onshore exploration wells provided by the Nigerian Geological Survey Agency (NGSA)/Nigerian Bitumen Corporation. From west to east (Fig. 3.2) the wells are Ilaro (IL), Ijagun (IJ), Akinmodiro (AM), Akinboni-Idogun (AK), Ohosu I (OH) and Ohosu II (SU). Core description files are provided in supplementary materials 3.1 A – F. In total, 46 samples from Ewekoro, Araromi and Afowo Formations, which are Cretaceous to Paleogene in age were collected for this study (9 limestone samples, 37 sandstone samples; Ewekoro Formation = 7 limestones, 5 sandstones; Araromi = 2 limestones, 17 sandstones; Afowo = 15 sandstone). The stratigraphic formations were assigned in each well based on the data from Omatsola and Adegoke (1981), Elueze and Nton (2004), Ikhane et al. (2011), Ogundana et al. (2014), Nton and Adeyemi (2015), Akinmosin et al. (2019), Adeigbe and Oyekola (2019), and Oluwajana et al. (2021).

Samples were impregnated with blue-dyed epoxy resin to highlight porosity and polished to a thickness of 30 µm for transmitted light microscopy. Ten of the 46 samples were prepared for further cathodoluminescence analysis. A mixture of Alizarin Red S (ARS; 6 grams) and potassium ferricyanide (PF; 4 grams) dissolved in 300 ml of 0.5% hydrochloric acid (HCl) was used for the staining to aid the identification of carbonate minerals. Petrographic analysis was performed using a Leica DM-LP transmitted-light microscope fitted with a Jenoptik Gryphax camera. Grain sizes were measured along the long axis of a minimum of 100 grains on a grid adjusted to the maximum observed grain size per sample (e.g. Busch *et al.,* 2018). Sorting was calculated after Folk and Ward (1957). Point-counting (300 counts) was conducted using a semi-automated Pelcon point counter mounted on a Leica DM-LP, with the step-length adjusted to the maximum grain size. Average point count and grain size results are given as arithmetic means.

Cathodoluminescence analysis was performed using a CITL MK5 stage fitted on a Leitz Aristomet microscope mounted with a Jenoptik Gryphax camera at an acceleration voltage of 10 – 20 kV (+/-2) and current of 325 – 350 μA.

The sandstone samples were classified based on a ternary plot of quartz-feldspar-rock fragments after McBride (1963). Samples rich in depositional clay mineral matrix (>15%) have been classified as wackes/matrix-supported sandstones. The intergranular volume (IGV %) indicating the degree of compaction was calculated from the point-count data and is equal to the sum of intergranular porosity and the pore-filling cements (Houseknecht, 1988). The classification of carbonate microfacies was done on the basis of composition and textural features (Dunham, 1962). Qualitative identification of the clay mineral was done by a standard X-ray diffraction method using Bruker D8™ diffractometer with a Cu cathode at 40 kV and 40 mA. No chemical pretreatment was performed. The semi-quantitative XRD measurements were analysed using Profex™ software and appropriate structural files to facilitate the identification of peaks (Doebelin & Kleeberg, 2015).

One inch (25.4 mm) cylindrical plugs were drilled parallel to bedding and used for petrophysical analyses. The porosity of the plugs (1-inch cylindrical samples; n=7 LST, 28 SST) was measured using Helium pycnometry (AccuPyc II 1340 by micromeritics®). Klinkenberg-corrected air permeability was measured at a constant confining pressure of 1.2 MPa using dry, oil-free lab air (80% N_2, 20% O_2) as the permeant (Monsees et al., 2020b). Microporosity was evaluated as the difference between He-porosity and point-counted porosity (optical porosity; Felder, 2011).

3.5 Results

3.5.1 Lithological description from core samples

The carbonate rocks of the Ewekoro Formation in the Ilaro Well (IL) are predominantly made up of wackestone (at 129 and 153 m) and packstone (between 128 – 138 m), which sometimes contain quartz sand (0.3 – 26.7% quartz), with interbedded fine- to medium-grained poorly sorted calcareous sandstones (between 136 to 158 m). The sandstones are often cemented by dolomite, contain glauconite and/or phosphate minerals. At the base, the limestone unit is sandy grading into the underlying Araromi Formation. The Araromi Formation is predominantly composed of quartz sandstones (between 162 – 201 m) which are sometimes calcareous and glauconitic, with interbedded shale (between 175 – 177 m) and clayey sandstone layers (between 178 – 179 m).

The clastic rocks of Afowo Formation in the Akinmodiro (AM) and Ijagun (IJ) wells share similar lithological characteristics consisting of quartz sandstones (between 52 – 54 m in AM) and clay-matrix supported sandstones (between 55 – 66 m in AM and at 91 – 107 m in IJ).

The studied section of the Akinboni (AK) well belongs to the Araromi Formation and consists of quartz sandstone (between 28 to 29.5 m; sometimes containing shell fragments) and a limestone layer of about 0.8 m thickness at 58.5 m depth. The interval between the analysed quartz sandstone and the limestone consists of unconsolidated sand.

The sediments at the top of the Ohosu I (OH) well belongs to the Araromi Formation and consist of shale (1 – 12.8 m) and limestone (12.8 – 13.8 m). The base consists of quartz sandstones (between 14 – 55.5 m) which belong to Afowo Formation and are often cemented by siderite.

Samples from the Ohosu II (SU) well belong to the Afowo Formation, which predominantly consists of fine- to medium-grained quartz sandstone. The top is ferruginized (at 28.5 m) and the base is siderite cemented (at 68 m).

In all the wells, unconsolidated sand (belonging to the Afowo and Araromi Formations) is present at some intervals. They are not considered in this study because they cannot be sampled intact.

3.5.2 Petrography

3.5.2.1 Depositional texture

The sandstones grain size mostly varies from very fine- to medium- (0.08 mm – 0.35 mm) (Fig. 3.4 a, b, Fig. 3.5) with the exception of one coarse-grained (0.77 mm) sandstone in the Ilaro well. The sandstones are generally poorly sorted (0.05 – 0.90). Point contacts are the most commonly observed textural characteristics in the siliciclastic rocks of Eastern Dahomey Basin (Fig. 3.4 a, b, c). The clay mineral matrix is detrital (Fig. 3.4 c, d), sometimes with the clay minerals deformed/bent around the grains which indicate the

presence of clay minerals prior to compaction.

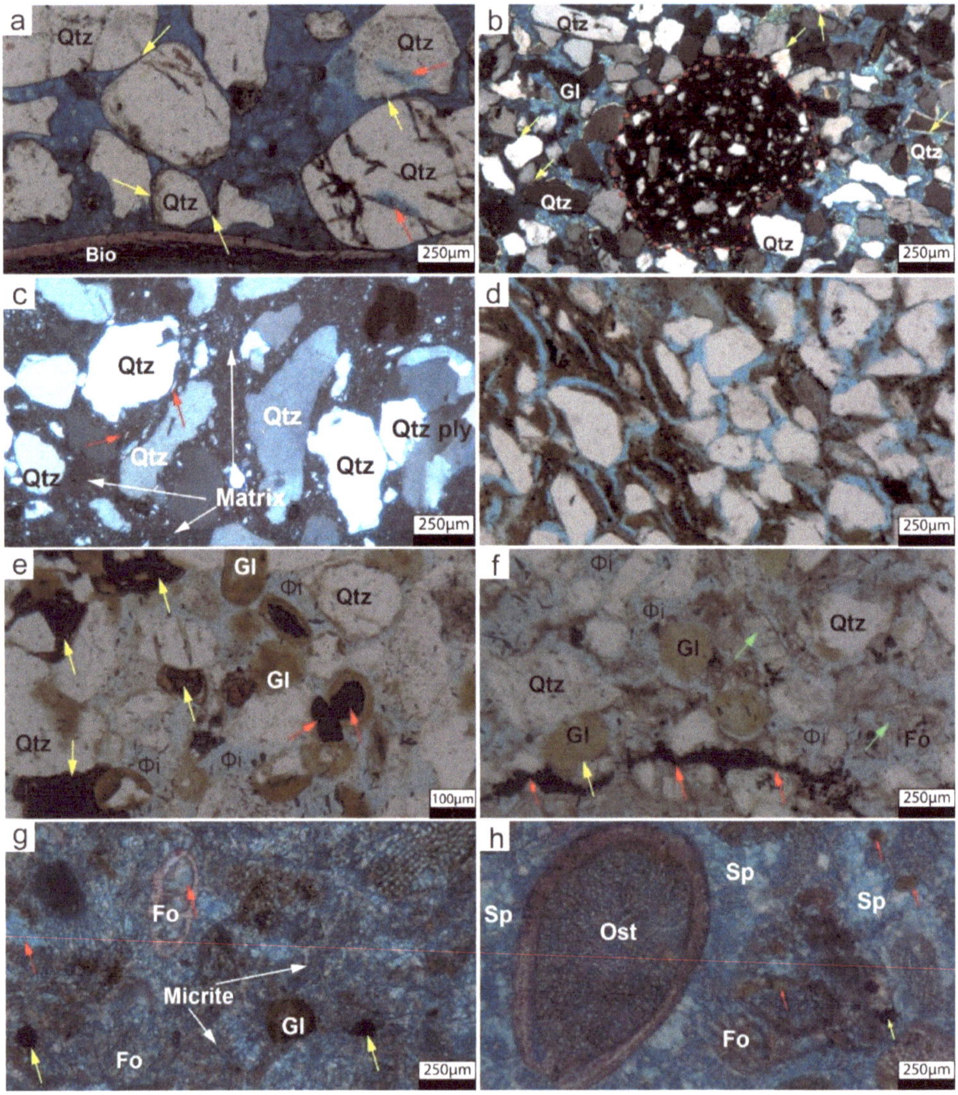

Figure 3.4. Figure caption on the next page.

Figure 3.4. Photomicrographs of typical textures and compositions of sandstones and limestones in the Dahomey Basin. a) Medium-grained sandstone (cf. Supplementary material 3.1 A) in the Ilaro well showing poikilotopic ferroan dolomite cement (stained blue), tangential/point contacts (yellow arrow) and inherited fractures (red arrow; sample no. IL 46; Araromi Formation; opt-Φ = 3.7%; porosity type: intergranular, secondary). b) Fine-grained sandstone (cf. Supplementary material 3.1 A) showing the effect of bioturbation (red dotted-circle). Grain contacts are tangential and point contacts (yellow arrow) with ferroan dolomite cement (stained blue; sample no. IL 47; Araromi Formation; opt-Φ = 5.0%; porosity type: intergranular, secondary) preventing long or sutured contacts. c) Clay matrix supported, very poorly sorted, medium-grained sandstone showing polycrystalline quartz grain (Qtz ply), monocrystalline quartz (Qtz), mica flakes (red arrow; sample no. IJ 07; Afowo Formation; opt-Φ = 1.3%; porosity type: intergranular, secondary). d) Laminated sandstone with detrital clay mineral laminae (sample no. IL 27; Araromi Formation; opt-Φ = 38.3%, porosity type: intergranular, secondary). e) Fine-grained glauconitic sandstone with phosphate grains (yellow arrow), opaque grains (red arrow), and intergranular pores (Φi; sample no. IL 39; Araromi Formation; opt-Φ = 30.7%; porosity type: intergranular, secondary). f) Glauconite rich sandstone with intergranular pores (Φi), secondary pores in shell fragments (green arrow) and pyrite along the laminae (red arrow). Qtz =quartz, Gl = glauconite, Fo = foraminifera, Foi = foraminifera inner lining, Φi = intergranular pores. g) Limestone with glauconite grains (Gl), pyrite (yellow arrow) foraminifera (Fo), and secondary pores in shell and cement (red arrow; sample no. IL 15; Ewekoro Formation; opt-Φ = 10.3%; porosity type: secondary). g) Limestone with sparite cement (Sp), ostracod (Ost), foraminifera (Fo), glauconite (red arrow) and pyrite (yellow arrow; sample number IL 17; Ewekoro Formation; opt-Φ = 1.0%; porosity type: secondary).

The limestones show a bioclastic texture embedded in micritic or sparry calcite cement (often dolomitic) surrounding or supporting skeletal grains, fossils, fossil fragments, quartz and glauconite grains (Fig. 3.4 e, 3.4 f). Micrite or infilling of voids by micritic mud and micrite precipitation in intraskeletal voids, and non-selective replacements of the original skeletal structures is evident in the shallow-marine limestones of both the Ewekoro and Araromi Formations in the Eastern Dahomey Basin, Nigeria (Fig. 3.4 g, h).

3.5.2.2 Detrital composition

3.5.2.2.1 Siliciclastic sandstones

Quartz is the dominant (23 – 63%) detrital component in all the sandstone samples and it occurs both as monocrystalline (Qm) and polycrystalline (Qp) grains, with the former being the most abundant (Supplementary material 3.2). The feldspar content of the samples is generally low (0 – 7.3%) and K-feldspar is more dominant than plagioclase. Plutonic and sedimentary (silty, and clay/shale) rock fragments (Rf) constitute 0 to 4.3% of the total framework. Hence, the analyzed sandstone samples are compositionally classified as quartz arenites, subarkoses (Fig. 3.6) and wackes/matrix-supported sandstone (McBride, 1963). Muscovite is the main mica present (0 – 3.7%) and is mainly restricted to the clay mineral matrix and is sometimes stained by iron oxide or oxyhydroxides (combined as FeOx). The clay mineral matrix content in the sandstones generally ranges from 0 to 34% with some of the wackes containing up to 70% clay mineral matrix. These consist of clay- and silt-sized mineral particles (< 2 – 63 µm) which are in some samples stained with iron oxide or are sometimes showing illitic/illite-smectite optical characteristics (especially in the Afowo Formation). Kaolinitic clay (avg. 14%, n=36, range: 0% – 70%), Illite (avg. 1.26%, n = 37, range 0% – 21.3%) and smectitic clays (avg. 6.6%, n = 37, range: 0% –

52.7%) are the main clay minerals present (mainly as detrital clay mineral matrix; Fig. 3.4 c). On occasion, the detrital clay mineral matrix also contains chamosite.

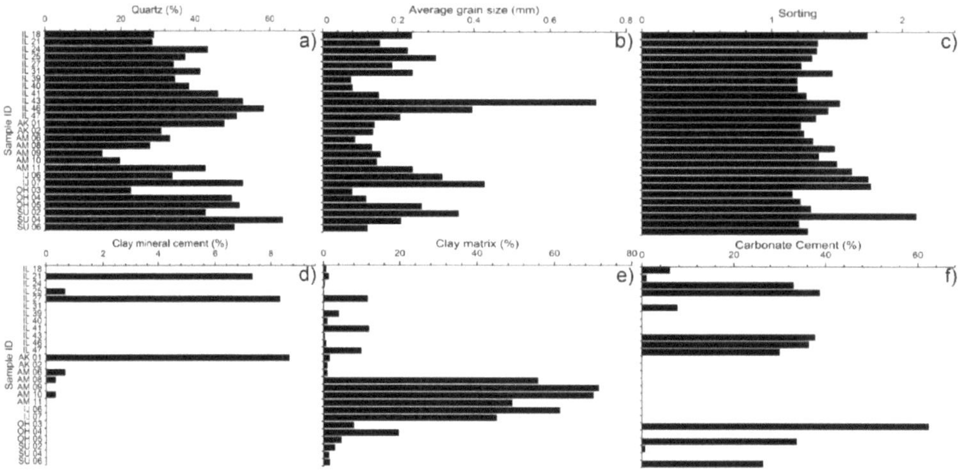

Figure 2.5. The bar charts showing variations in quartz, average grain size, sorting, clay mineral cement, clay mineral matrix, and carbonate cement in the analysed samples.

Accessory heavy minerals in the samples are mainly zircon (0% – 0.7%), rutile (trace), tourmaline (0% – 0.3%) and very few opaque mineral grains (0% – 0.7%). Shell and fossil fragments of various forms and compositions are encountered in most of the samples (0% – 8%; Fig. 3.4 a). However, these are rare in the clay-matrix supported samples. The sandstones sometimes contain phosphatic grains (Fig. 3.4 g, h).

3.5.2.2.2 Limestones

The limestones consist mainly of allochems (mainly fossil fragments) and in some instances quartz and glauconite grains (Supplementary material 3.3; Fig. 3.4 e, f). Shells of foraminifera, pelecypods, brachiopods, echinoids, ostracods and other undifferentiated fragments are often encountered in the limestone of the Ewekoro and Araromi Formations (Fig. 3.4 g, h).

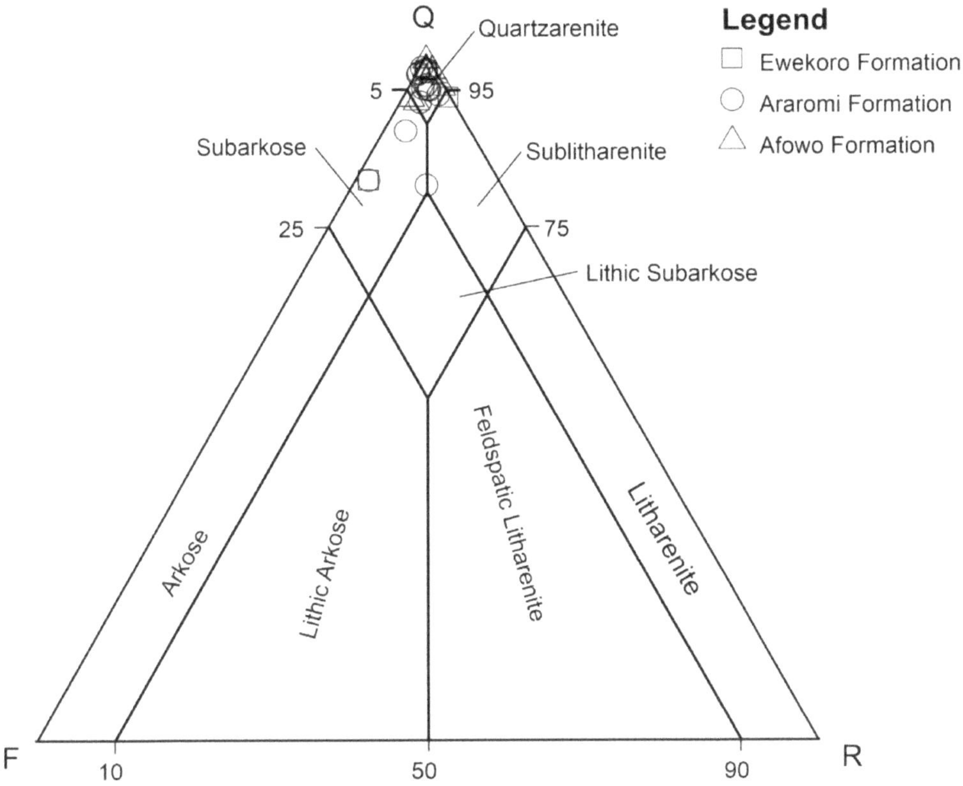

Figure 3.6. Ternary plot of quartz (Q), feldspar (F) and rock fragments (R) of the siliciclastic sandstones showing the quartzarenite to subarkosic composition (after McBride, 1963).

3.5.2.3 Authigenic phases

3.5.2.3.1 Clay minerals in the sandstone

The sandstones of Afowo Formation contain some authigenic kaolinite (as feldspar replacement, avg. 0.3%, n = 15, range: 0% – 1.33%; Fig. 3.7 a, c, d). The kaolinite occurs in vermiform booklets and in most cases is present within intragranular pores or completely replaced detrital feldspar grains. Only a trace amount of illite is also found replacing kaolinite in some samples. The dolomite- and calcite-cemented sandstone samples usually have low detrital and authigenic clay contents.

3.5.2.3.2 Carbonate cements

3.5.2.3.2.1 Sandstones

Among the 37 sandstone analysed, 13 samples contain carbonate minerals. Ferroan dolomite (stained blue), ferroan calcite (stained mauve), and siderite (unstained) are the main observed carbonate cements. Only one sample (IL 46) also contained non-ferroan dolomite (0.7%) in addition to ferroan-dolomite.

Pore-filling Fe-dolomite and Fe-calcite often show a poikilotopic or rhombohedral texture encasing the detrital constituents (Fig. 3.4 a, b; Fig. 3.7 b, e – i). Unlike the Fe-calcite and Fe-dolomite, siderite rarely forms as extensive pore-filling cement but occur mostly as discrete fine crystals within the pores (Fig. 3.7 e, f). Ferroan-dolomite and ferroan-calcite are the dominant type of carbonate cement occurring in the Ewekoro and Araromi Formations, ranging from 0 – 36.7% (avg. 3.6%, n = 37) and 0 – 37.7% (avg. 1.5%; Supplementary material 3.2) respectively. Siderite occurs only in the Afowo Formation in wells Ohosu I and II ranging from 0 up to 62% (Fig. 3.7 e, f). Some areas do not contain carbonate cements when filled by detrital clay mineral matrix (Figs. 3.4 d, 3.4 e). On few occasions, the carbonate cemented sandstones are bioturbated with visible burrow that contains finer-grained sediments (Fig. 3.4 b).

3.5.2.3.2.2 Limestones

The bioclasts in the limestones are cemented with Fe-dolomite and/or Fe-calcite or non-ferroan dolomite. These grains are embedded in a groundmass of micrite (lime mud) and/or sparite cement (Fig. 3.4 g, h, Fig. 3.7 h, i, j).

3.5.2.3.3 Dissolution

3.5.2.3.3.1 Sandstones

Intragranular pores are present in rock- and fossil fragments. The most observed altered fragments are feldspar grains and shell fragments, which are sometimes partially or wholly dissolved resulting in intragranular and mouldic pores in the clay mineral matrix of the sandstones (Fig. 3.7 c, d).

3.5.2.3.3.2 Limestones

Some bioclasts are dissolved in the limestones, leading to vuggy pores (Fig. 3.4 e). These unfilled pores resulting from the previous dissolution of shells contribute to the overall porosity of the limestone (Fig. 3.4 g, h).

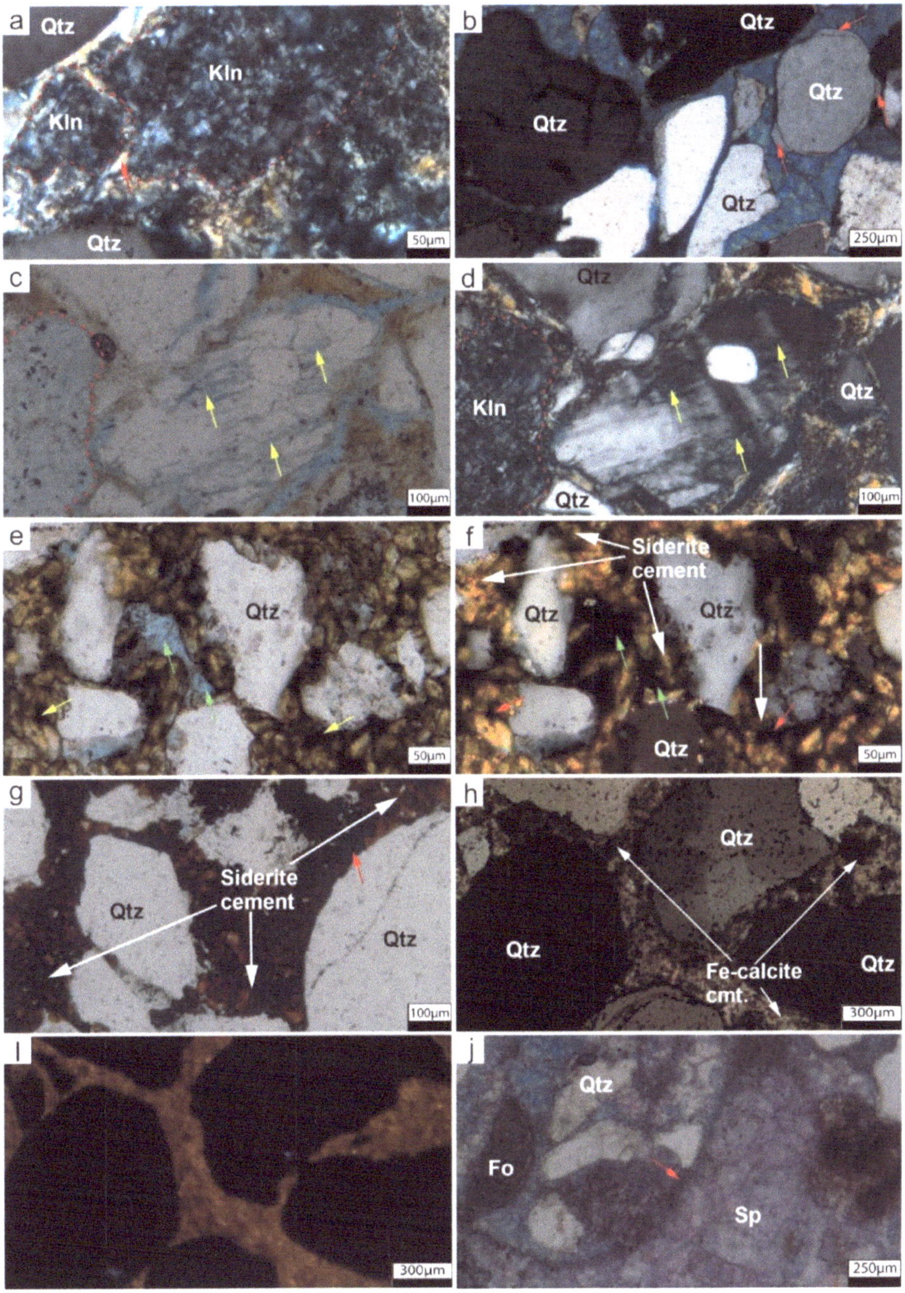

Figure 3.7. Figure caption on the next page.

Figure 3.7. Photomicrographs of authigenic components under plane polarized light (ppl) and cross polarized light (xpl). a) Kaolinite with micropores, likely replacing a feldspar grain. The former grain outline is still visible (red broke line) in a detrital clay matrix rich sandstone (xpl; sample no. IJ 08; Afowo Formation; opt-Φ = 5.3%; porosity type: intergranular, secondary). b) Fe-dolomite cemented sandstone with inherited quartz cement (see rounded abraded outlines; red arrow; sample no. IL 46; Araromi Formation; opt -Φ = 3.7%; porosity type: intergranular, secondary). c, d) Ppl and xpl image showing secondary pores in feldspar suggesting dissolution after deposition (yellow arrow,), kaolinite with micropores (red broken line; sample no. IJ O8; Afowo Formation; opt-Φ = 5.3%; porosity type: intergranular, secondary). e, f) Siderite cemented sandstone of the Afowo Formation (e=ppl, f=xpl) with little clay matrix and pyrite (red arrow). Green arrows indicate the intergranular pores (sample no. OH 04; Afowo Formation; opt-Φ = 28.7%; porosity type: intergranular, secondary). g) Siderite cemented sandstone with iron oxide (FeOx) stain (dark brown) and pyrite (yellow arrow; sample no. OH 05; Afowo Formation; opt-Φ = 8.7%; porosity type: intergranular, secondary). h, i) The cathodoluminescence (CL) image of dolomite cemented sandstone showing orange luminescence colour of the dolomite cement within the pore space (xpl; sample no. IL 43; Araromi Formation; opt-Φ = 5.7%; porosity type: intergranular, secondary). j) Packstone showing foraminifera test (Fo), detrital quartz grains (Qtz) and sparite cement filling a skeletal element (ppl; sample no.IL 19; Ewekoro Formation; opt-Φ = 0.0%; porosity type: no optical porosity). opt-Φ = optical porosity.

3.5.2.3.4 Non-binding and rarely cementing authigenic minerals

3.5.2.3.4.1 Sandstones

Pyrite (avg. 0.7%; n = 37; range: 0 – 4.3%;) occurs in most of the samples in all formations. It often occurs in framboidal crystal clusters or is occurring in between the laminae or mouldic/secondary pores of shells (Fig. 3.4 e – h).

Granules and pellets of green clays (interpreted to be derived from the glauconite facies) are commonly present in sandstones of the Ewekoro and Araromi Formations (avg. 3.9%; n=22; range: 0% – 20.7%) whereas no occurrence is recorded in the clayey sandstone of the Afowo Formation. The term glauconite is employed in this study to broadly indicate the greenish clay granules. Glauconite is mostly encountered in this study as lobate pellet and less frequently deformed and squeezed between more rigid detrital grains or as grain coatings of quartz and opaque mineral grains (Fig. 3.4 e, f). Glauconite also appears encased in carbonate cements (Fe-dolomite in Fig. 3.4 b) and in association with pyrite (Fig. 3.4 f).

Phosphates (0 – 5.6%; avg. 0.6%; n = 36) mostly occur as small distinct grains, bound to the surface of other grains or on occasions within pores (Fig. 3.4 e). Where phosphate grains are embedded in clay minerals, the surrounding clay minerals appear with an orange colour alteration. This can be observed in sample SU 02 where 26% of the total framework belong to this group (Fig. 3.4 i).

3.5.2.3.4.2 Limestones

Glauconite is present in wackestones and packstones of the Ewekoro Formation (Fig. 3.4 e, f; 0% – 20.7%; avg. 4.4%; n = 7) and exhibits the same characteristic as in the sandstone.

3.5.2.3.5 Compaction and intergranular volume

The intergranular volume in the sandstone ranges from 7.7% to 63.0% (avg. 30.81, n=28). Argillaceous and micaceous grains in sandstones show weak bending and stretching. Their long axis is aligned parallel to bedding. Glauconite grains also show some deformation (Fig. 3.4 h) in the calcareous sandstones. Point contacts (Fig. 3.4 a, b, c) are prominent in some of the samples while others are loosely bound by cement or clay mineral matrix. Overall, mechanical compaction in the studied samples is small.

3.5.3 Mineralogy

X-ray diffraction analysis indicates the presence of quartz in bulk rock and <63 microns' separate analyses (39% and 14% respectively; Fig. 3.4 a – f; Fig. 3.7 a – i; Fig. 3.8 a, b). Smectite (18%), kaolinite (38%) are the main clay mineral phases identified in the clay-matrix supported sandstone of Afowo Formation (Fig. 3.8 a). The clay minerals in the Araromi and Ewekoro Formations consist dominantly of smectite (44%) and kaolinite (38%; Fig. 3.8 b). Other minerals present are glauconite (2%), chamosite (9%) and pyrite (Fig. 3.8 b). The limestone of the Ewekoro Formation is mineralogically made up of calcite (38%) dolomite (20%) and ankerite (35%), with subordinate kaolinite (2%), quartz (2%) and glauconite (2%; Fig 3.8 c).

3.5.4 Optical porosity

Porosity observed in thin section analysis is classified as intergranular porosity (0% – 32.3%; avg. 13.2%; n = 37; Fig. 3.4 d – f, Fig. 3.7 e, f), intragranular porosity (0% – 11%; avg. 2.9%; n = 37; Fig. 3.4 e, h, Fig. 3.7 c, d) and microporosity (Fig. 3.4 c, d, Fig. 3.7 a, e, f). The intragranular porosity resulted from grain dissolution (Fig. 3.7 c, d), shell fragment dissolution (Fig. 3.4 f - h), and authigenesis of kaolinite replacing feldspars (Fig. 3.7 a) while microporosity is only associated with carbonate cement and partly authigenic clay minerals (Fig. 3.4 c, e, f, Fig. 3.7 a, e, f). These micropores exist between small crystals of siderite (Fig. 3.7 e – g), clay mineral matrix and carbonate mud/micrite (Fig. 3.4 a, b, Fig. 3.7 a, c, d). The optical porosity obtained from the clastic rocks of the Ewekoro Formation ranges from 2.7 – 31.7% (avg. 19.5%, n = 5) while it ranges from 1.0% – 38.3% (avg. 20.5, n = 17) and 1.3% – 32.7% (avg. 10.1%, n = 15) in the Araromi and Afowo Formations respectively (Supplementary material 3.2).

Contributing to the total optical porosity in the limestone is shell dissolution, resulting in vuggy and mouldic pore types (secondary porosity, Fig. 3.4 g, h). Limestones in the Ewekoro Formation have optical porosity values ranging from 0% – 10.3% (avg. 3.0%,

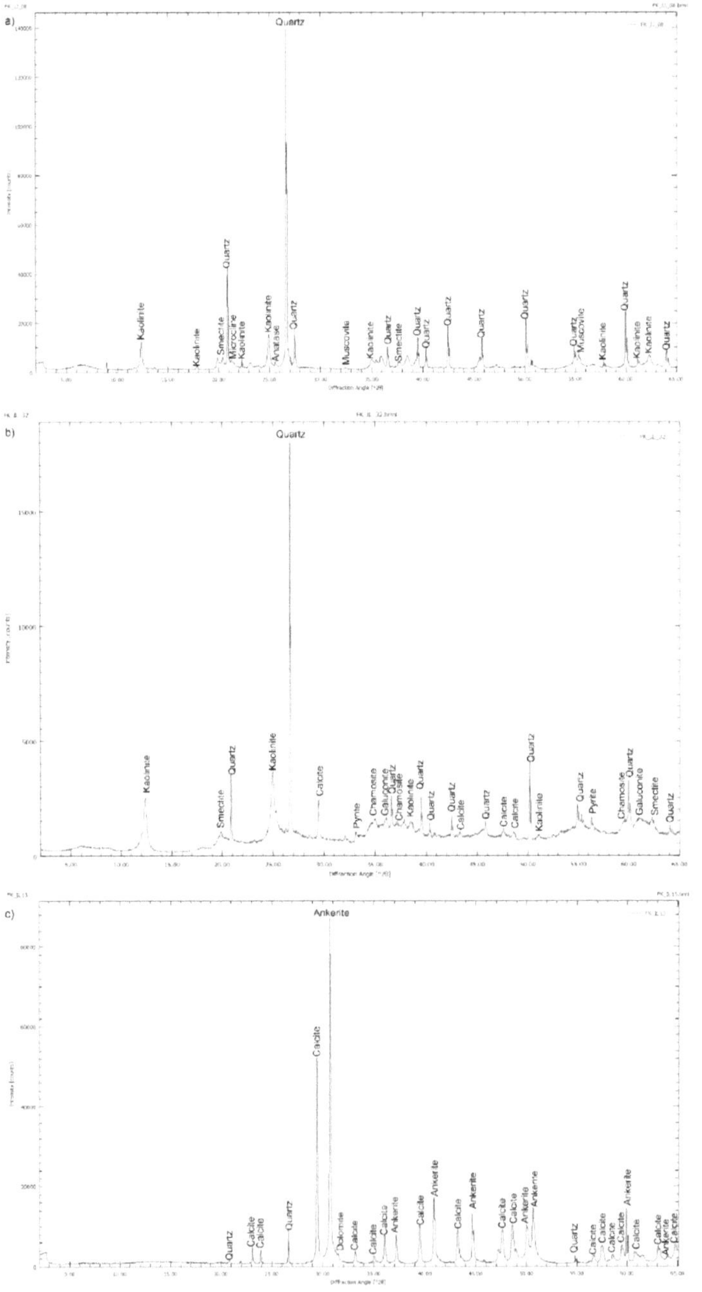

Figure 3.8. X-ray diffractogram showing the identified mineral phases. a) Clay matrix rich sandstone of Afowo Formation (IJ 08; whole rock). b) glauconitic sandstone of Araromi Formation (IL32; <63 microns separate). c) Limestone of Ewekoro Formation (IL13; whole rock).

n = 7) while the value obtained in the limestone of Araromi Formation are 0% and 10% (avg. 0.5, n = 2).

3.5.5 Petrophysical data

He-porosity values in the clastic rocks of the Ewekoro Formation ranges from 5.4% – 35.8% (avg. 19.7%, n = 4), and it ranges from 4.3% – 45.2% (avg. 28.1, n = 10) and 14.6% – 46.6% (avg. 30.4%, n = 14) in the Araromi and Afowo Formations, respectively (Supplementary material 3.2). Sandstones of the Ewekoro Formation exhibit permeability ranging from 0.006 – 109.2 mD (avg. 29 mD, n = 4). In the Araromi and Afowo Formations, permeability of the sandstones varies from <0.001 – 1764 mD (avg. 402 mD, n = 10) and 0.05 – >10000 mD (avg. 1451 mD, n = 14) respectively.

The limestones of the Ewekoro Formation have porosity in between 6.7% and 31.7% (avg. 15.9%, n = 6) and the only limestone of the Afowo Formation has a porosity of 8.0% (Supplementary material 3.3). Permeability in the Ewekoro limestones range from 0.0037 – 12.9031 mD (avg. 2.6 mD, n = 6) while the limestone in the Araromi Formation has a permeability of 0.03 mD (Supplementary material 3.3).

As expected, the measured He-porosity in each of the samples is higher than that observed in thin section (Supplementary material 3.2, supplementary material 3.3, Fig. 3.9 a). The differences in the measured porosity (He-porosity) and the optically counted macroporosity is interpreted to have resulted from the contribution of microporosity. The highest amount of microporosity is found in the clay-rich sandstone/clay-matrix supported sandstone. Pore spaces between micron-sized crystals of the micrite are the most volumetrically significant type of microporosity in the limestone.

3.5.6 Porosity and permeability distribution

Samples from the Afowo Formation (in AM and IJ) containing more clay mineral matrix (>20%) have higher He-porosity at increasing clay mineral matrix contents (Fig. 3.9 b). No consistent correlation is observed from Ewekoro, Araromi and Afowo Formations in IL, AK, OH, and SU wells. He-porosity is reduced in the carbonate cemented sandstones while the permeability falls within the same range as the clay–matrix supported sandstone (Fig. 3.9 c, d). Generally, the permeability is lower in samples with lower IGV values, as opposed to samples at the same porosity but higher intergranular volume (IGV; Fig. 3.8 c). The matrix-supported sandstones (wackes) have generally lower permeability than quartz-rich sandstones (Fig. 3.8 e). For both subgroups porosity and permeability are positively correlated. A negative correlation exists between grain size and porosity. The finer-grained sediments have relatively higher porosity (Fig. 3.9 f).

The cross plot of porosity (%) and grain density (g/cm^3) shows an increase in porosity with increasing grain density and a positive correlation exists between porosity and permeability in the limestone (Fig. 3.10 a, 3.10 b).

3.6 Discussion

3.6.1 Lithofacies and environmental interpretation

The clastic facies of the Ewekoro and Araromi Formations consist of fine to very fine-grained, poorly sorted glauconitic sandstones that are often carbonate cemented and calcareous. Glauconite is a prominent autochthonous mineral in shallow-marine environments (Odin & Matter, 1981). It is the green to dark green, Fe-rich, mica-type clay minerals of marine origin and with K_2O ranging from 2% to > 6% (Huggett, 2013; Odin & Matter, 1981). The autochthonous character of the glauconite in the studied samples is supported by the grain coatings around detrital grains (Fig. 3.4 e, f). This coupled with the occurrence of a number of shell fragments indicate deposition in a shallow-marine setting, probably between 10 to >500 m water depth (Odin & Matter, 1981) on the shelf or slope environment. In modern depositional settings, glauconite minerals are known to occur in water depths greater than 50 m and are common in shelf and slope environments (O'Brien et al., 1990; Odin & Matter, 1981). The fossil assemblages in the limestone of the Ewekoro and Araromi Formations consist of foraminifera, brachiopod, ostracod, gastropod and other undifferentiated shell fragments. This also indicates deposition in shallow-marine settings.

The sandstone of the Afowo Formation consists of quartz-rich and clay-matrix supported (wacke) types. The depositional texture of the quartz sandstone and the presence of bioturbation (burrows; Fig. 3.4 b) indicate deposition in shallow-marine environments. The intervals containing poorly consolidated sand are due to poor lithification of the sediments in the subsurface. Similar observations have been reported in the Niger Delta which makes conventional cores difficult to obtain (Iboyi & Odedede, 2014; Lambert-Aikhionbare, 1981; Lambert-Aikhionbare, 1982). The composition of the quartz-rich sandstones is generally comparable to the sandstone of Agbada Formation in the Niger Delta Basin. However, the amount of early diagenetic carbonate cement and glauconite in the Ewekoro and Araromi Formations of the Dahomey Basin is presumably higher than that of the deeply buried Niger Delta sediments (Lambert-Aikhionbare, 1981).

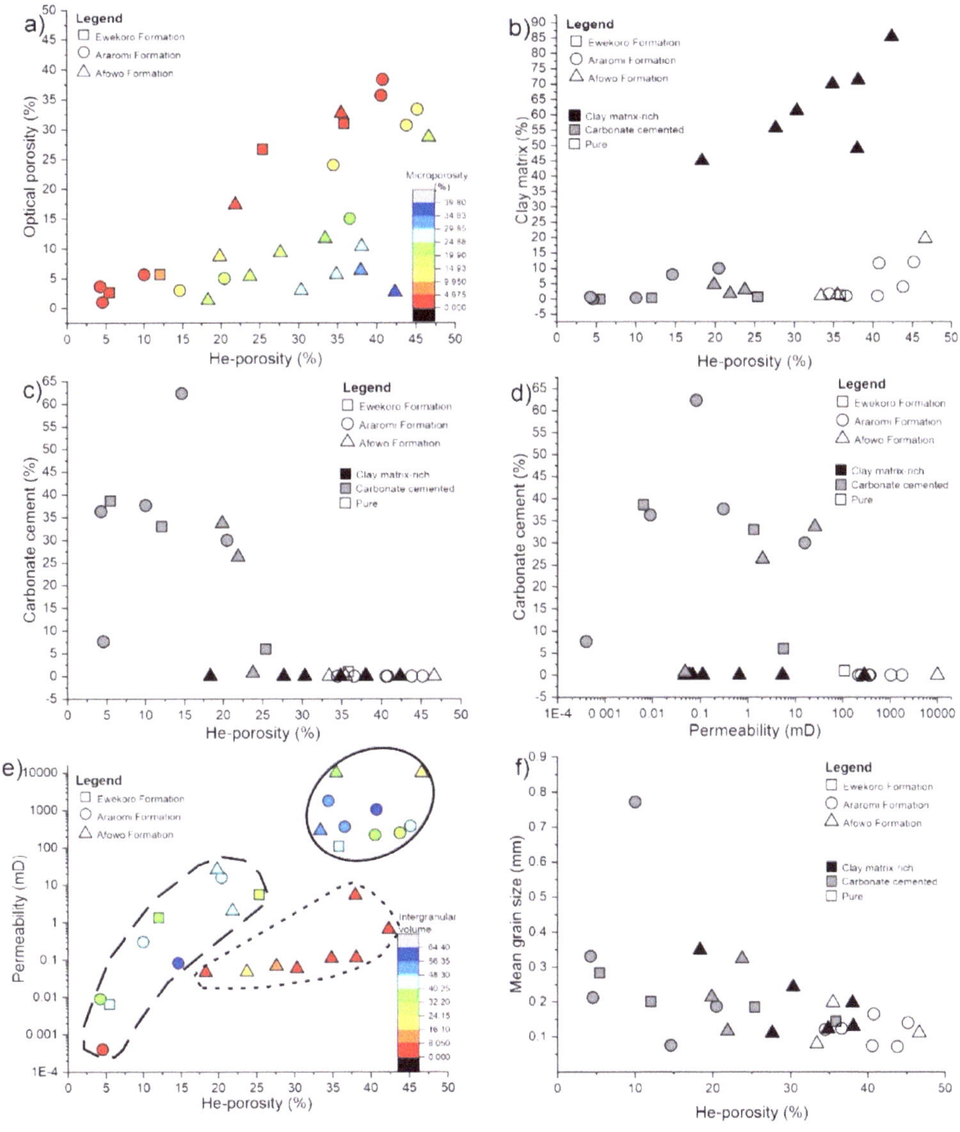

Figure 3.9. Figure caption on the next page.

Figure 3.9 a) He-porosity versus optical porosity showing no correlation. Microporosity is higher in the clay mineral matrix rich sandstone of Afowo Formation in AM and IJ wells. b) He-Porosity versus clay mineral matrix content of the sandstone, indicating the significant contribution of the micropores to the measured He-porosity in the clay mineral matrix-rich sandstones (filled black). The carbonate cemented samples (filled grey) have lower porosity compared to the pure quartz sandstones (not filled). c) Carbonate cement versus He-porosity showing that carbonate cemented samples (filled grey) have lower porosity compared to the rest of the samples. d) Carbonate cement versus permeability showing lower values in the carbonate cemented and clay-matrix supported sandstone samples. e) He-porosity and permeability showing positive correlation in the porosity-permeability relationship. Quartz sandstone with no clay mineral matrix and no carbonate cement (solid line) have higher permeability than the remaining populations of carbonate cemented (long dash line) and clay-matrix rich (short dash) sandstones. The clay matrix-supported sandstones show high porosity some of which are comparable to the samples containing less clay mineral matrix, yet permeability is much lower because of unconnected micropores. f) Porosity versus grain size generally showing weak negative correlation, with the higher porosity recorded in the finer-grained sediments.

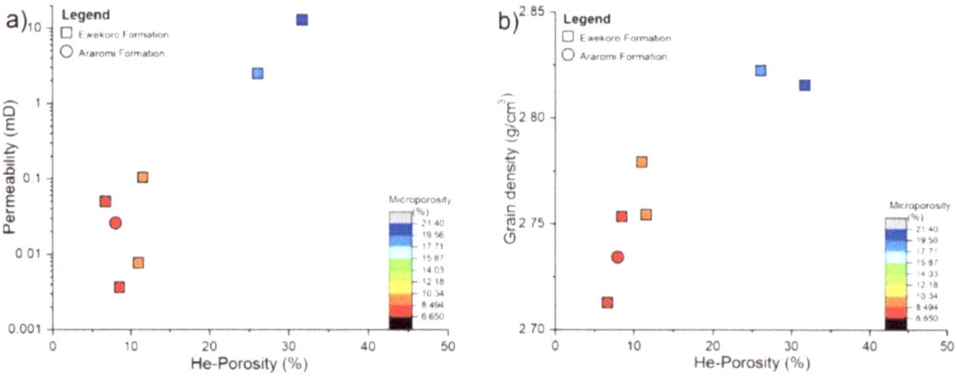

Figure 3.10. a) Porosity versus permeability cross plot of the limestone samples showing a positive correlation. b) Porosity versus grain density plot. Dolomite cement accounts for the high density. The micropores within the sparite cement (sparry) account for increasing porosity with density.

3.6.2 Paragenetic sequence

Based on the published burial reconstruction for the Afowo-1 well at the coastline west of Lagos, the post-Turonian sedimentary rocks in the basin might have not at any point in time been subjected to temperatures greater than 75°C (Onuoha & Ofoegbu, 1988). Phases of uplift are not recorded (Onuoha & Ofoegbu, 1988). The fact that sediment loading rather than deep-seated tectonic events is responsible for subsidence in the upper part of the Afowo-1 well (Onuoha & Ofoegbu, 1988) allowed a deduction that the wells studied in the present work, comprising Upper Mesozoic and younger lithologies represent their maximum burial depths at the present day. Based on this interpretation, the studied samples in the onshore region are still in the early diagenetic realm (eodiagenesis/shallow burial).

The textural relations obtained from thin section analyses are used to assess the relative timing of diagenetic processes. The most common diagenetic features observed in the sedimentary rocks of the Eastern Dahomey Basin, besides minor compaction, are carbonate cementation (Fig. 3.4 a, b, I; Fig. 3.7 b, e – I), authigenic pyrite formation (Fig. 3.4 e, f), glauconite formation (Fig. 3.4 b, e – h), feldspar dissolution and replacement by clay minerals (Fig. 3.7 a, c, d).

The relative paragenetic sequence of dominant diagenetic features are systematically placed as early and late in these overall shallowly buried siliciclastic and carbonate rocks (Figs. 3.11, 3.12).

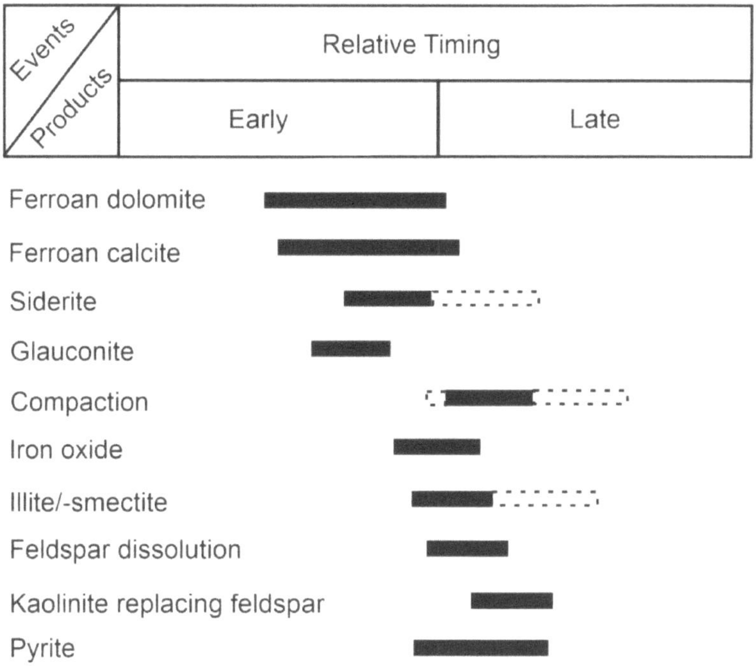

Figure 3.11. Paragenetic sequence of diagenetic processes within the Mesozoic – Cenozoic siliciclastic rocks of the eastern Dahomey Basin.

3.6.2.1 Glauconite

The earliest diagenetic phase is the in-situ formation of glauconite grains in a shallow marine environment in a humid climate (Huggett et al., 2015). As glauconite is encased in carbonate cements (Fig. 3.4 b), thus glauconite must have been present prior to carbonate cementation or been formed contemporaneously. The occurrence as coatings on detrital quartz and mineral grains additionally supports the formation following initial sediment deposition. The immature granules indicate in situ formation because they are soft and not

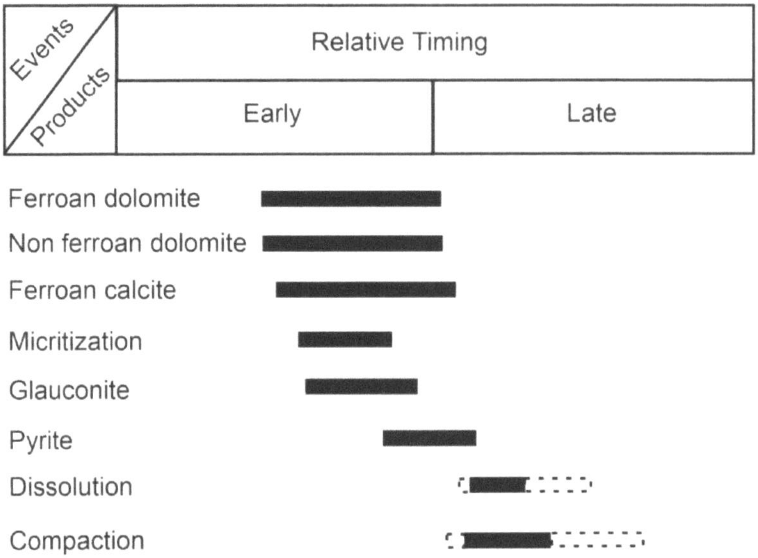

Figure 3.12. Paragenetic sequence of diagenetic processes within the Mesozoic – Cenozoic carbonate rocks of the eastern Dahomey Basin.

likely to survive transportation more than a few meters without breaking up (Fig. 3.4 g, h; Huggett, 2013). The formation of glauconite close to the sediment-water interface indicates the presence of Fe, K and Al in the solution from which glauconitic minerals could be formed (Huggett, 2013). These elements are believed to be sourced from hinterland weathering of the Precambrian Basement Complex of Southwestern Nigeria (lithogenic input) and are deposited into the ocean mainly via river discharge, with little or no mineral dust contribution. The observed paragenesis with pyrite (Fig. 3.4 f) is also described to be a common association in the shallow marine realm (Kelly & Webb, 1999).

3.6.2.2 Carbonate cementation

The bulk of the carbonate cement (Fe-calcite and Fe-dolomite) is interpreted to have formed during early marine diagenesis. Apparently, it prevented compaction (Fig. 3.4 a, b) as indicated by the high IGV and the lack of long or sutured contacts between detrital grains. Moreover, floating grains and only point contacts of grains encased in Fe-dolomite and Fe-calcite cement (Fig. 3.4 a, b, Fig. 3.7 b, h, i) indicate an early diagenetic formation prior to compaction. Sea/pore water, biogenic carbonate and carbonate intraclasts are likely the carbonate source for carbonate cement precipitation (Curtis & Coleman, 1986; Naiman,

1982). The presence of glauconite likely formed in-situ indicates a fairly Fe- and Mg-rich fluid from which precipitation of ferroan dolomite/calcite could occur (Curtis & Coleman, 1986; Estupiñan et al., 2007).

Though still within the early diagenetic realm, siderite is interpreted to represent a relatively later phase of carbonate cementation. The occurrence of siderite cement is restricted to some of the relatively older sediments of the Afowo Formation (Fig. 3.7 e - g, Fig. 3.12) and there is no observed paragenetic interaction of dolomite or calcite with siderite in a single sample (Supplementary material 3.2). Micropores (Fig. 3.7 e – g) between the tiny crystals of siderite maintained some porosity of the rocks in these horizons and additionally inhibited later porosity loss by compaction (Dutton et al., 1996).

In the Niger Delta Basin, east of the Dahomey Basin, calcite and siderite cements are also present and the solution phase transformation (where calcite is first dissolved and later replaced by siderite) was favored (Lambert-Aikhionbare, 1982). Siderite in the present study, however, is interpreted to have been precipitated from anoxic Fe-rich pore water during early diagenesis in a brackish depositional environment that existed in some part of the Afowo Formation of the Dahomey Basin.

This brackish environment has been reported by Omatsola and Adegoke (1981) and the transitional/brackish to marginal marine conditions were said to represent a littoral or estuarine near-shore depositional environment in which the water level fluctuated (Omatsola & Adegoke, 1981). Brackish conditions for siderite precipitation are favored in this study because of the absence of glauconite and other full marine indicators in the siderite-cemented horizons. Siderite formation has been reported by Bahrig (1994) to be characterized by a high Fe/Ca ratio in pore water which favors siderite precipitation rather than calcite in an at least periodically stagnant water column. Thus, in the Eastern Dahomey Basin, Nigeria, the content of Ca/Fe ions in solution is low enough to permit the precipitation of siderite within the older sediments of Afowo as opposed to calcite/dolomite precipitation in the younger sediments of Araromi and Ewekoro Formations due to differing environmental conditions. Hence, the Fe/Ca ratio, pH/ Eh condition, solubility, and input of CO_3^{2-} from organic matter maturation are among the major factors controlling different types of carbonate precipitation (Lambert-Aikhionbare, 1982; Morad, 1998; Pearce et al., 2013; Postma, 1981).

3.6.2.3 Early micrite formation

Syndepositional micrite or infilling of voids by micritic mud and micrite precipitation in intraskeletal voids, and non-selective replacements of the original skeletal structures is evident in the shallow-marine limestones of both the Ewekoro and Araromi Formations in the Eastern Dahomey Basin, Nigeria (Fig. 3.4 g, h). Oluwajana et al. (2020) recognized some diagenetic micritization of bioclasts in wackestones of the Benin Flank of the Dahomey Basin. This intrabasinal process leads to precipitation and replacement of initially deposited shell fragments by micrite (Fig. 3.4 g, h). Micritization is common in a marine environment (Khan et al., 2019; Ojo, 2004). In this study, early ferroan and non-ferroan dolomite formation precedes micritization while sediment compaction followed progressively (Fig. 3.4 a, b; Fig. 3.7 b, h, I; Fig. 3.11; Fig. 3.13).

3.6.2.4 Quartz cementation and clay minerals

Authigenic quartz cement is absent in the studied samples of the Eastern Dahomey Basin. Only some abraded and rounded syntaxial quartz overgrowths are present on detrital grains, visible by a dust rim separating it from the detrital grains, which are interpreted to be inherited (Fig. 3.7 b).

Lambert-Aikhionbare (1982) identified a few quartz cements in sandstone samples of the Agbada Formation of the Niger Delta Basin. This is to some extent comparable to the observed lack of quartz cementation in the sandstones of the Ewekoro, Araromi and Afowo Formations in the Eastern Dahomey Basin. The low quartz cementation indicates a low degree of chemical compaction supplying silica in solution (Bjørlykke, 1998; Bjørlykke & Jahren, 2010). The absence could also be related to limited clay mineral (especially smectite) recrystallization, releasing silica into solution (McKinley et al., 2003). Quartz precipitation kinetics are driven by temperatures of about 70°C or higher (Busch et al., 2018b; Walderhaug, 2000) and affected by clay mineral coatings (McBride, 1989; Molenaar & Felder, 2018; Monsees et al., 2020b).

Authigenic clay minerals are also rare in the studied samples. However, minor amounts of kaolinite are found to be replacing feldspar (Fig. 3.7 a, c, d) in some of the samples. Minor dissolution also occurs in some feldspar grains (Fig. 3.7 c, d). The K-feldspar to kaolinite transformation is a common early diagenetic alteration either during meteoric water flushing or in relation to CO_2-rich or organic acid rich fluids in association to meteoric waters (Bjørkum & N., 1988). Illitic optical properties (bright yellowish to golden colour birefringence) is observed within the clay matrix, but the matrix is interpreted to be largely detrital (Fig. 3.4 c, d, Fig. 3.7 a, c, d). The primary composition of it is unknown, no conclusion could be made about the diagenetic influence/modification as the primary material could be made up of a mixture of kaolinite, illite and smectite or illite-smectite mixed-layer clay in varying proportions. This makes the quantification of authigenic and detrital clay a very complex endeavor as the clay mineral matrix and the authigenic clay are not distinguishable in the XRD result and occurs as heterogeneous mixture in thin section.

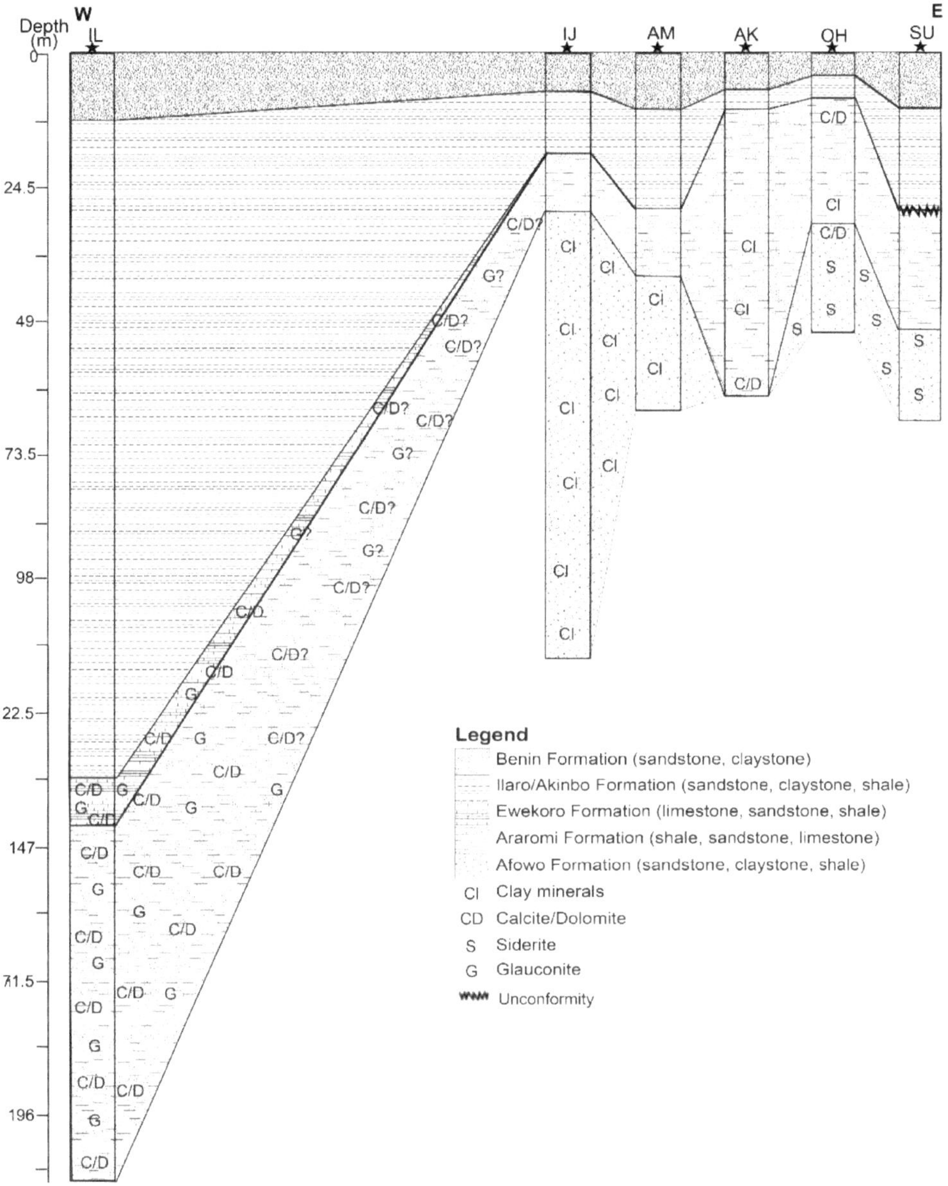

Figure 3.13. Correlation of the wells with overlay of dominant diagenetic minerals. The occurrence of siderite is restricted to Afowo Formation while dolomite and calcite occur in both Ewekoro and Araromi Formation.

3.6.2.5 Compaction

The detrital clay mineral matrix shows some degree of compaction, as the detrital clay mineral matrix is deformed around rigid quartz grains (Fig. 3.4 d, e). The effect of overburden pressure is reduced by early carbonate cementation and the clay mineral matrix, as evident from IGV calculations. This in turn inhibits compaction in the sandstones and limestones (Fig. 3.4 a, b; Fig. 3.7 b, h, i). Greater porosity loss could occur in the glauconitic sandstones as a result of the enhanced compaction due to the deformation of ductile grains (Schulz-Rojahn et al., 2003). However, the predominance of highly stable quartz grains and the presence of early diagenetic carbonate cement is expected to reduce the overall effect of porosity loss resulting from the deformation of ductile grains. The shape of the glauconite pellets/granules could also be useful in first-hand reservoir quality assessment, especially in the fine-grained sandstones at greater depths where relatively greater deformation is expected except being inhibited by other diagenetic processes (Schulz-Rojahn et al., 2003).

3.6.3 Reservoir quality controls

The reservoir quality of the studied sandstone generally varied widely from fair to excellent (porosity from 4.3% to 46.6%, permeability from 0.0004 to >10,000 mD; Fig. 3.14 a – c; Tiab and Donaldson, 2012). The clay-matrix supported sandstones (wackes) in Afowo Formation represent a net reducing/poor reservoir quality due to their generally low permeability (which is not expected to improve at depth) despite high porosity. Detrital clay mineral matrix has a reducing influence on the overall reservoir quality of the sandstone (Fig. 3.9 b, e). Thus, the high porosity in the matrix-rich sandstone (18.3% – 42.3%) is expected to be reduced at greater depths due to the effect of compaction and the corresponding loss of micropores. These clay-rich and poor reservoir quality samples are typical of some deposits within the Afowo Formation, which are found in the Ijagun (IJ, Fig. 3.4 c) and Akinmodiro (AM) wells.

Close packing of the sediments was prevented by the early diagenetic cementation as evident by the IGV (Fig. 3.9 e) which indicate a reduced effect of compaction on the studied samples. However, the carbonate cementation reduces the available intergranular pore spaces (Fig. 3.14 b). Development of intragranular porosity and micropores (especially in siderite cemented sandstone of the Afowo Formation, Fig. 3.7 e, f) contributes to slightly improved porosity (8 to 79% of the total porosity). This is in agreement with studies by Karim et al. (2010) on Lower Cretaceous reservoir sandstones in the western Sable Subbasin, offshore Nova Scotia. A lesser amount of early diagenetic carbonate cement is expected to be present in sediments deposited further offshore and burial diagenetic recrystallization of carbonate cements may be sparse. This is expected to improve the reservoir quality of the deposits further offshore.

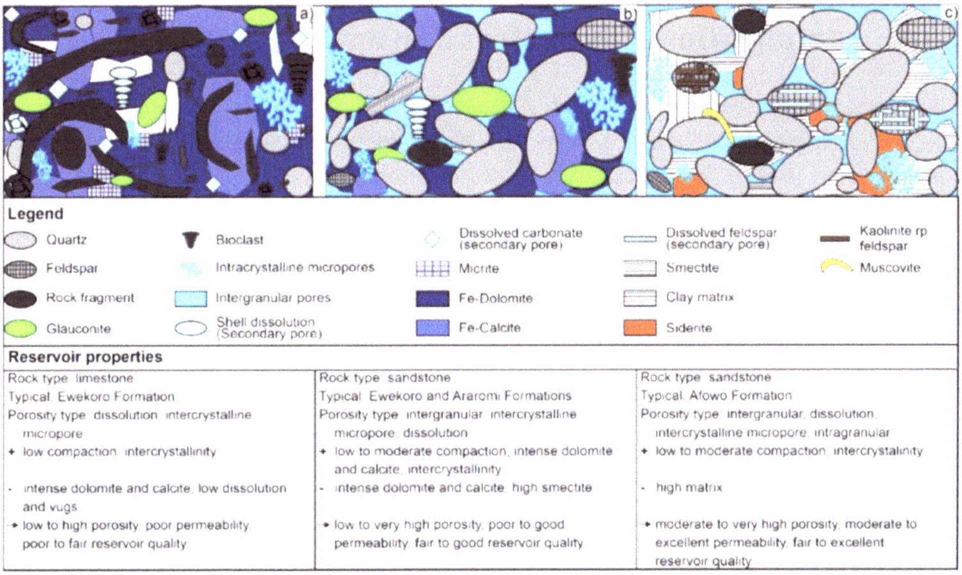

Figure 3.14. Schematic micrographs of typical limestone and sandstones showing different compositional frameworks, porosity types, porosity and permeability class, and reservoir quality. a) limestone of Ewekoro Formation. b) basal sandstone of Ewekoro, and Araromi Formations. c) clay-matrix rich sandstone of Afowo Formation.

Cross-plots of petrophysical and petrographic data were used to highlight reservoir quality controls (Fig. 3.9 a – f; see section 3.6.7). While the effect of microporosity is relatively more significant in the total porosity value (Fig. 3.9 a), the corresponding permeability is less significant (except for some samples with well-connected pore networks; Fig. 3.9 e). A better porosity-permeability relationship exists in the quartz sandstones than the clay matrix-supported counterparts (Fig. 3.9 e). The clay mineral matrix, including the abundance of micropores together with micropores within the carbonate cement accounts for fairly higher porosity in finer-grained sediments (Fig. 3.9 d).

Calcite/dolomitic cement and micrite account for the high density in the limestone (Supplementary material 3.2 B). Additionally, the grain density correlates with He-porosity, implying that a higher density carbonate mineral, i.e. dolomite correlates with higher porosity (Fig. 3.10 a, b). Compaction and cementation must have led to some level of reduction in the porosity (Adekeye & Akande, 2006; Adekeye et al., 2005). However, the secondary porosity resulting from the dissolution of some of the bioclast and initial calcite/dolomite cement, and the micropores within the micrite keeps the porosity fairly good. The corresponding low/poor permeability indicates the presence of largely unconnected pores (Fig. 3.14 a – c).

3.6.4 Implications for exploration and exploitation

High porosity in the sandstones of Afowo Formation coupled with their minimal mechanical compaction resulting from early cementation and high clay mineral matrix content could lead to the production of fines and formation damage during production. However, due to the dominance of kaolinitic clay minerals in the matrix and quartz in the detrital framework, the effect of compaction resulting from overburden pressure is expected to be smaller compared to smectite rich sandstone and/or sandstones with high percentages of feldspar and rock fragments. Production of fines ordinarily could pose a major engineering problem in a basin (Lambert-Aikhionbare & Shaw, 1982). However, it could also help to maintain permeability within a reservoir if the fines are brought to the surface, depending on the type of clay, and drilling fluid used (Almon & Davies, 1978; Lambert-Aikhionbare, 1981; Lambert-Aikhionbare & Shaw, 1982). Kaolinite does not react, only smectitic clay could react with drilling fluids and introduced fluid during enhanced production.

The clay mineral matrix rich facies of the Afowo Formation is of concern because of their low permeability compared to their arenitic counterpart. However, the permeability might not possibly be reduced much further during production because of the dominance of largely kaolinite group clay minerals assemblages. With kaolinite as the main clay mineral, permeability loss is minimized with suitable reservoir management and development plan (Akinmosin et al., 2019; Busch et al., 2018b; Lambert-Aikhionbare, 1981; Lambert-Aikhionbare, 1982). Hence, the presence of kaolinite/dickite clay minerals in the quartz sandstone of the Afowo Formation is expected not to have significantly reduced the reservoir quality or negatively impact enhanced recovery operations. The problem with kaolinite is easily overcome through the use of available clay stabilization systems (Almon & Davies, 1978).

The engineering problem induced by the clay mineral assemblages may be further complicated in the Dahomey Basin by the presence of smectitic clay in the Ewekoro and Afowo Formations. These may swell in contact with the injected water, thereby causing permeability loss and corresponding formation damage (Almon & Davies, 1978; Lambert-Aikhionbare & Shaw, 1982). It is therefore necessary to carefully select the drilling fluid while penetrating the smectite bearing horizons in the Ewekoro and Araromi Formations. With an oil-based fluid or potassium chloride drilling, completion and stimulation systems interference by smectite swelling can be avoided.

More transformation of a largely kaolinitic clay mineral assemblage (Fig. 3.8 a, b) to illite and illite-smectite mixed-layer clay is expected in deeper horizons as formation temperature increases (Iboyi & Odedede, 2014; Lambert-Aikhionbare, 1981; McKinley et al., 2003; Odigi, 1994). But that is not the case in the deeply buried sediments of the Niger Delta where smectite was found in excess of what is expected from published models of clay mineral diagenesis (Lambert-Aikhionbare & Shaw, 1982; Odigi, 1994). This is believed to have resulted from undercompaction resulting from rapid deposition (Lambert-Aikhionbare & Shaw, 1982). In the studied shallow onshore wells in the Dahomey Basin, weakly cemented and uncompacted sandstones occurred at some intervals as well as non-tar bearing outcrops of the Afowo Formation. The probable presence of this in the deeper sections of the Dahomey Basin especially with the low silica cementation calls for special caution as migration of fines and swelling of clays are liable to result in various hydrocarbon recovery and engineering problems.

3.7 Conclusions

The sedimentary rocks of the Eastern Dahomey Basin recovered from shallow onshore exploration wells show a low level of compaction and lithification. Some intervals were rarely cemented due to the absence of cementing materials during and/or after deposition. Early diagenetic carbonate cementation in the shallow-marine sediments of the Eastern Dahomey Basin reduced much of the original intergranular porosity, thereby stabilizing the framework and inhibiting compaction. Depending on the amount of carbonate cement ranging from 0 to 39%, some interparticle porosity is preserved. The clay mineral matrix has a net negative effect on the reservoir properties of the sandstones. The micropores within the matrix contribute significantly to the porosity, but the corresponding permeability values in the matrix-rich sandstone samples is significantly low (0.05 to 5.3 mD).

Good reservoir quality is established in the shallow-marine sandstones in the Afowo and Araromi Formations (porosity from 14.6 to 47% and 5 to 36% respectively), the former represents the best potential reservoir in this study and should be targeted for further exploration. An exception is given by the facies containing clay mineral matrix which is deposited in estuarine settings. Although the limestones exhibit good porosity values (7 to 32%), the permeability is fairly low (0.003 to 13 mD) due to the vuggy and mouldic pores as a result of dissolution, creating individual disconnected pores. Therefore, fair to good reservoir properties is assigned to the limestones. The sedimentary deposits of the Eastern Dahomey Basin are highly heterogeneous in terms of lithological, stratigraphic, diagenetic and petrophysical properties. In future exploration, it is therefore necessary to avoid overgeneralization of any factor in targeting reservoir rocks for oil and gas exploitation.

4 Diagenetic evolution of continental to shallow marine Mesozoic–Cenozoic deposits, Eastern Dahomey Basin, Nigeria: Implications for reservoir quality

4.1 Abstract

The Eastern Dahomey Basin is a continental to shallow marine hydrocarbon-bearing basin formed as a marginal pull-apart basin during the Late Mesozoic and Cenozoic. The heterogeneity of the sedimentary deposits filling the basin makes accurate location and prediction of reservoir rocks and their properties difficult. This difficulty is further complicated by the poor understanding of diagenetic processes and timing, and their influence on porosity. A proper understanding of the heterogeneity of diagenetic events promises a better assessment of the reservoir quality evolution. Sampling of outcrop analogues and petrographic analyses allow the identification of detrital and authigenic components, as well as pore types, and allow to delineate reservoir quality controls.

Outcrop and shallow well samples from the Cretaceous (Ise, Afowo, Cretaceous Benin Flank, Araromi Formations) and Paleogene (Ewekoro, Akinbo-Oshosun, Ilaro Formations) were retrieved from the surface or a maximum depth of 35.4 m and generally exhibit poor to excellent optical porosity (2.7% to 39.7%; avg. 19.4%; $n = 66$) for the siliciclastic rocks of the basin. Porosity distribution is largely controlled by the clay mineral matrix content and Fe-mineral precipitation. The main pore types are intergranular and intragranular. The sampled near-surface siliciclastic sediments are within the early diagenetic realm where only iron oxide (FeOx) and iron oxyhydroxide (FeO(OH)) cementation, minor kaolinite and pyrite precipitation, and feldspar and possible carbonate dissolution constitute the major diagenetic modifications. Furthermore, the occurrence of oil sands indicates that these shallow reservoir rocks have locally been charged. In comparison to shallow subsurface samples from the region, intense FeOx and FeO(OH) cementation is only observed in outcrops, while subsurface samples contain Fe-rich carbonate cements. Porosity reduction other than that associated with compaction (at greater depth) is predicted to be low unless carbonate cements are still preserved.

Poor to fair optical porosity is present in the limestones, with values ranging from 0 to 6.3% (avg. 2.6%, n = 11). The wackestone and packstone samples have the highest optical porosity ranging from 0.3 to 6.3% (avg. 4.2%, n = 6). Sparry carbonate cements and lime mud in bioclastic wackestones and packstones reduced the available interparticle porosity (to 0%). The cementstones only contain negligible optical porosity while sparstones retain some porosity in intercrystalline pores between dolomite rhombs (0 to 0.7%).

4.2 Introduction

Fundamental to the success of oil and gas, geothermal, and any storage capacity exploration is the accurate location of reservoir rocks and the assessment of their quality. This involves the evaluation of rock composition, provenance, diagenetic alteration, porosity, permeability, and structural framework. Generally, the quality of a reservoir rock is primarily measured by the porosity and permeability. These are dependent on the pre-depositional (e.g. source rock lithology, weathering and transport), and post-depositional factors (e.g. cementation, dissolution and compaction). The source dependent detrital composition of sandstones also has a great influence on the overall reservoir quality as it affects the diagenetic pathway physically, chemically, or both (Bloch, 1994; Morad et al., 2010; Rossi et al., 2002). The composition of detrital grains, intensity of weathering (mechanical or chemical) and grain size are strongly determined by the lithology of rocks in the catchment area (Arribas & Tortosa, 2003; Caracciolo, 2020; Hessler & Lowe, 2006; Palomares & Arribas, 1993). Climate and the transport medium also affects the composition of sediments deposited (Johnsson, 1993; Mack, 1978), as unstable grains as feldspar could be partly or totally dissolved leaving intragranular or mouldic porosity. This is because, the reactivity of each mineral differs in each transporting medium and climate.

Furthermore, the source rock composition and transport are important in the overall depositional characteristics of sedimentary rocks as they control the grain size distribution, degree of sorting, and roundness of sediments. These factors, in combination with the cement and matrix content, are further important factors that control primary porosity in reservoir rocks (Adamolekun et al., 2022; Busch et al., 2018b; Monsees et al., 2020a), where good sorting and large grain sizes often favor better primary porosity in clastic sediments.

Post-depositional factors including cementation, compaction, and dissolution affect the amount of remaining porosity in siliciclastic rocks. The original primary porosity may be destroyed by cement precipitation, while grain and cement dissolution, resulting in secondary porosity, may enhance porosity. The effect of compaction is usually evident in the amount of intergranular volume (IGV) and grain contact types in sediments

(Houseknecht, 1988; Paxton et al., 2002). The effect of mechanical compaction may be reduced by mineral precipitation in which the grain framework is stabilized by authigenic cements (Busch et al., 2022; Monsees et al., 2021; Schmidt et al., 2020a). This may lead to improved reservoir quality where high amount of intergranular porosity still exist, unless mineral precipitation is pervasive.

Often, the type and quantity of cementing materials have a direct influence on reservoir quality (Bjørlykke, 2015; Boyd et al., 2004; Monsees et al., 2020a). For instance, cementation by poikilotopic calcite could completely occlude the intergranular porosity while the drusy siderite crystals may retain some intergranular and intercrystalline porosity. Furthermore, authigenic clay minerals (e.g. kaolinite, illite) may either form pore filling or pore lining cements. While the pore lining cement pose very little obstruction to the intergranular porosity, the pore filling cements partly or completely occlude the intergranular porosity in reservoir rocks. The sequence of cementation and/or the timing is the most crucial factor when investigating reservoir quality.

Petrographic studies have helped to deduce basin evolution and tectonics, petrofacies, paleoclimate, paleo-drainage, sediment dispersal patterns, and improved stratigraphic correlation which are important in exploration (Dickinson & Suczek, 1979; Garzanti, 2019; Garzanti et al., 2016; Ingersoll, 1978; Nton & Adamolekun, 2018; Tobin & Schwarzer, 2014). In addition, quantitative petrographic analyses of siliciclastic rocks have proven useful in estimating porosity in the absence of data from direct rock-plug measurement by fluid injection (Anovitz & Cole, 2015; Makeen et al., 2016). Limestone petrography also provides useful information on its composition, texture, microfacies, compaction and porosity types from which reservoir quality can be reliably assessed (Oluwajana et al., 2019; Saller et al., 2016; Scholle & Ulmer-Scholle, 2003).

Data and literature on the quality of reservoir rocks in Eastern Dahomey Basin in Nigeria is limited. The impact of oil emplacement on the diagenesis in Cretaceous oil sands has been discussed by Bata et al. (2017), who estimated the average porosity in the oil sand of Afowo Formation at the time of emplacement to be 17.8%. Based on the analysis of samples from shallow exploration wells, Adamolekun et al. (2022) identified carbonate cement and clay mineral matrix as the main controls on reservoir quality of siliciclastic rocks in the basin. The reservoir quality of the carbonate rocks was identified to be controlled by dissolution in forming secondary porosity. Reservoir potential of the carbonate rocks have been reported to be very low by other authors based on petrographic studies (Adekeye & Akande, 2006; Adekeye et al., 2005; Oluwajana et al., 2020). In this paper, we utilized petrographic methods, especially the evaluation of diagenetic processes to assess the porosity evolution and reservoir potential of the Cretaceous to Paleogene fluvial to shallow marine siliciclastic and carbonate rocks in the onshore Eastern Dahomey Basin. The applicability of outcrop analogues to assess subsurface reservoir quality in the same formations within the region is also discussed with reference to data in Adamolekun et al., 2022. This study will provide valuable insight into the porosity evolution, distribution and control in the basin and aid future exploration attempts.

4.3 Geological setting

The Dahomey Basin, a marginal pull-apart basin covering approximately 32,000 km^2, was initiated in the Late Mesozoic during the opening of the South Atlantic Ocean resulting from the separation of the African and South American plates following the break-up of Gondwana (McLoughlin, 2001; Omatsola & Adegoke, 1981; Fig. 4.1). This marginal pull-apart basin extends from the eastern limit of the Volta Delta (Ghana) in the west through Togo and the Benin Republic to the western limit of the Niger Delta (Nigeria) in the east. The Nigerian sector of the Dahomey Basin is often referred to as the Eastern Dahomey Basin. Sedimentation in the basin started in the Lower Cretaceous and continues to present day.

The sedimentary deposits of the Eastern Dahomey Basin range from clastics (sandstone, siltstone, shale, claystone) to carbonates (limestone) with variable thicknesses and compositions (Fig. 4.3). Late Mesozoic to Early Cenozoic sandstones, siltstones, and limestones constitute the potential reservoir rocks in the basin. Generally, they were deposited in lacustrine, fluvial, and marginal marine to shallow marine and marine environments (Akinmosin et al., 2019; Doust, 1990; Nton & Adeyemi, 2015; Omatsola & Adegoke, 1981).

Due to the limited outcrop exposures and well coverage, the stratigraphic nomenclature of the Eastern Dahomey Basin is debated. The most accurate and widely accepted stratigraphic interpretation of Omatsola and Adegoke (1981) is adopted in this study (Fig. 4.3). The oldest sedimentary deposits in the Eastern Dahomey Basin lie unconformably on the crystalline basement rocks of South-western Nigeria and are composed of conglomerates, coarse- to medium-grained sandstones with intercalations of claystone (Valanginian/Barremian to Albian Ise Formation (IS); Omatsola & Adegoke, 1981). This is interpreted as a pre-drift sequence, deposited in a fluvio-lacustrine environment (Omatsola & Adegoke, 1981). The Ise Formation is overlain by medium- to fine-grained sandstone with inter-bedded siltstone and shales (Turonian to Maastrichtian Afowo Formation (AF); Nton & Otoba, 2011). The fluvial to littoral/brackish or near-shore sandstones are sometimes oil-bearing or oil-impregnated (Billman, 1976, 1992; Omatsola & Adegoke, 1981).

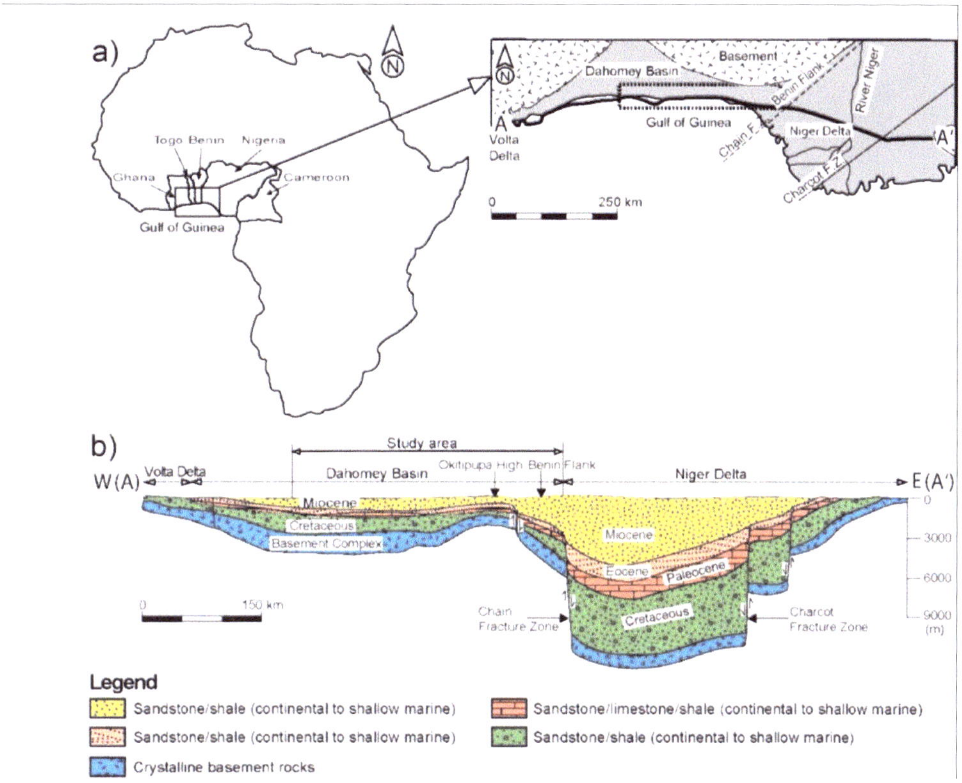

Figure 4.1. a) Map of Africa showing the location of the Dahomey Basin and the study area (black dotted rectangle; see Fig. 4.2). b) The west-east cross-section (line A- A' in Fig. 4.1 a) of Nigerian coastal basins; Dahomey and the Niger Delta (from Ikhane et al., 2011, after Whiteman, 1982b).

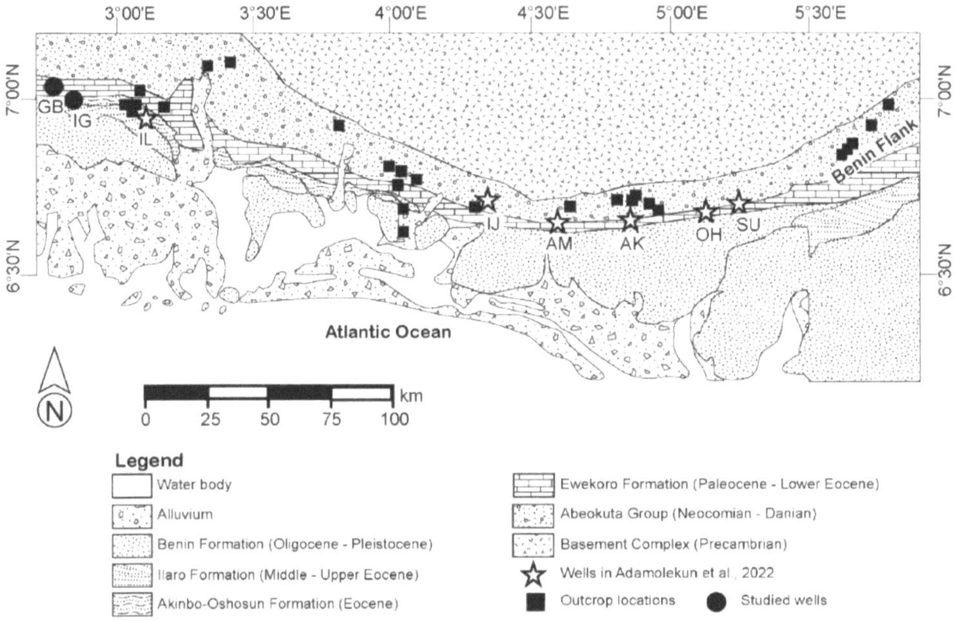

Figure 4.2. Onshore geological map of Eastern Dahomey Basin showing the locations of the studied outcrops and wells in the study area indicated by the black dotted rectangle in Fig. 4.1 (after Billman, 1976; Omatsola and Adegoke, 1981).GB = Gbokoto, IG = Igbeme, Il = Ilaro, IJ = Ijagun, AM = Akinmodiro, AK = Akinboni, OH = Ohosu I, SU = Ohosu II.

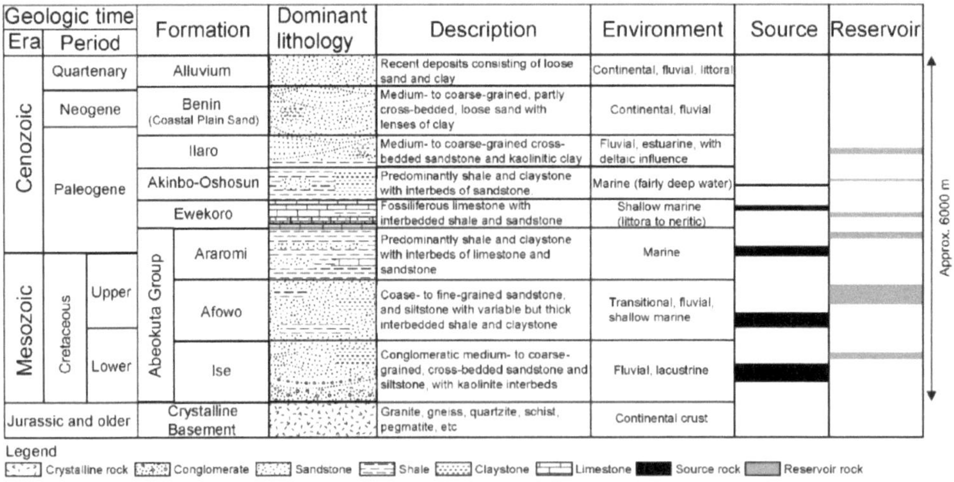

Figure 4.3. Simplified stratigraphic column of the Eastern Dahomey Basin showing stratigraphic and lithological units. Lithological symbols according to United States Geological Survey (USGS) standards (after Agagu, 1985; Billman, 1976; Omatsola and Adegoke, 1981).

The youngest formation within the Cretaceous consists predominantly of shale, and interbedded siltstone, sandstone, and limestone. (Maastrichtian to Paleocene Araromi Formation (AR); Omatsola & Adegoke, 1981). Deposition took place in a shallow marine to deep marine environment. Marl and lignite are also present within this formation in some localities.

A local classification of the Cretaceous deposits is often used for the easternmost part of the basin mostly referred to as the Benin Flank (Fig. 4.1; Fig. 4.4; Omatsola & Adegoke, 1981). In this study, the occurring deposits are simply referred to as Cretaceous Benin Flank (CBF) deposits and are further subdivided based on their prominent grain size. The coarse-grained, medium-grained and fine-grained sandstone in this region are assigned CBF-C, CBF-M and CBF-F respectively. The sedimentary rocks of the Cretaceous Benin Flank consist of medium- to coarse-grained conglomeratic cross-bedded sandstone (CBF-C), ferruginous sandstone, fine- to medium-grained sandstone (CBF-M), fine-grained sandstone, limestone, shale, and/or claystone (CBF-F). The CBF-C, CBF-M and CBF-F are similar to the Cretaceous lithologies in the western flank of the basin;(Ise, Afowo, and Araromi Formations).

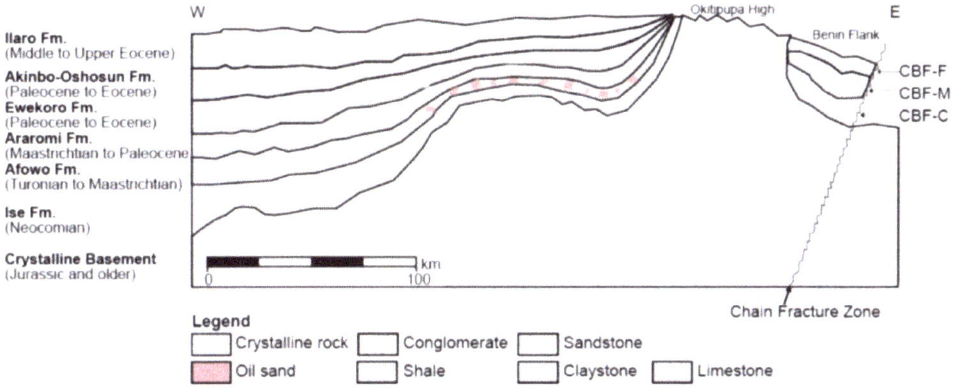

Figure 4.4. Schematic diagram of lithological distribution in the Eastern Dahomey Basin. The lithology of the Benin Flank is similar to the Cretaceous deposits of the western part of the Eastern Dahomey Basin. CBF-C = coarse-grained sediment in the Cretaceous Benin Flank, CBF-M = medium-grained sediment in the Cretaceous Benin Flank, CBF-F = fine-grained sediment in the Cretaceous Benin Flank.

The oldest Cenozoic unit is predominantly composed of limestone, with interbedded shale and sandstone (Paleocene to Eocene Ewekoro Formation (EW); Nton, 2001; Okosun, 1998). The abundance of marine fossils (coralline algae, gastropod, pelecypod, echinoid, etc.) in this formation is indicative of shallow marine deposition (Fayose, 1970; Nton, 2001; Whiteman, 1982c). It is overlain by a grey shale and/or claystone sequence and greenish-grey or beige clay and shale with interbeds of sandstone (Paleocene to Eocene

Akinbo and Ypresian Oshosun Formation (AK-OS)). The shale is glauconitic, thickly laminated and phosphate-bearing (Nton, 2001; Okosun, 1998).

The overlying formation consists of thick, massive, yellowish to brownish, poorly sorted, cross-bedded sandstone (Middle to Upper Eocene Ilaro Formation (IL); Nton, 2001; Omatsola and Adegoke, 1981). This formation is dominantly continental with varying estuarine, deltaic and marine influence (Adediran et al., 1991; Olabode & Mohammed, 2016).

The youngest sedimentary deposit in the Eastern Dahomey Basin consists of poorly sorted, non-fossiliferous, partly cross-bedded sands with lenses of clays. It is inferred to have been deposited in a transition to continental environments (Neogene Coastal Plain Sand of the Benin Formation; Billman 1976; Okosun, 1998). This formation is capped by the recent alluvium deposits of unconsolidated sand and clay (Adamolekun & Umoh, 2016; Falufosi & Osinowo, 2021; Olabode & Mohammed, 2016).

4.4 Materials and Methods

4.4.1 Outcrop studies

The Ise, Afowo, Araromi, Ewekoro, Akinbo-Oshosun (Akin.-Osh.) and Ilaro Formations from the Eastern Dahomey Basin and the Cretaceous equivalents from the Benin Flank (named as CBF-C, CBF-M and CBF-F) have been investigated in this study (Table 4.1). Twenty-four outcrop sections of the Eastern Dahomey Basin, including the Cretaceous Benin Flank (CBF), exposed in South-western Nigeria were studied using field observations and petrographic methods (Fig. 4.2, supplementary material 4.1 for detailed locations). In addition, the limestone of the Ewekoro Formation recovered from two shallow exploration boreholes (Gbokoto (GB) and Igbeme (IG)) in the Eastern Dahomey Basin were also studied, with a TVD of 35.4 and 34.2 m respectively (Fig. 4.2, supplementary material 4.1). At each outcrop, we described the lithology, colour, texture, and sedimentary structures.

4.4.2 Thin section petrography

A total of 77 thin sections consisting of 66 sandstones (Ise = 11, Afowo = 24, CBF = 11, Ilaro = 16, Akinbo-Oshosun = 4) and 11 limestone (Araromi = 2, CBF = 1, Ewekoro = 8) samples were prepared for this study (supplementary material 4.2, 4.3). The samples were impregnated with blue-dyed epoxy resin to highlight porosity and prepared to a thickness

Table 4.1. Lithology and descriptions of the studied formations in the Dahomey Basin.

Formations	Lithologies	Localities	Descriptions
Ilaro	Sandstone, claystone, shale	Ilaro, Ajegunle	Fine- to coarse-grained, yellowish to brownish sandstone (with clay clasts which are most times dissolved) and interbedded claystone. The sandstone is sometimes cross-bedded.
Akinbo-Oshosun	Shale, claystone, sandstone	Odolewu-Ijebu, Ilaro, Augustine University area	Light- to dark-grey fissile shale and claystone with subordinate light grey to brownish, matrix-supported sandstone.
Ewekoro	Limestone, claystone, shale	Ibeshe, Gbokoto, Igbeme	At Ibeshe quarry, the limestone is fossiliferous, with interbedded shale. At the top of the two shallow wells, it consists of fissile greyish shale and claystone below which are interbedded limestones. The limestone is light to dark-grey and contains very few visible fossils, vugs and vein.
Araromi	Shale, limestone	Ayadi	Light- to dark-grey shale. Interbeds of grey to dark grey, fossiliferous limestone (with visible shells) is sometimes present in the eastern part.
Afowo	Sandstone, shale	Ijebu-Imusin, Mbolufon and Ijebu-Ife, Looda, Orisunbare-Ilubinrin, Orita-J4, Egbe	Predominantly composed of fine-to medium-grained (occasionally coarse-grained) sandstone with interbedded shale. The sandstone is whitish to yellowish, generally finning upward, friable/unconsolidated to weakly consolidated, with parallel bedding, and occasionally crossbedding. Ferruginous sandstone layers are also locally present. The formation contains solid hydrocarbon in some outcrops (oil-sand) where it ranges from dark-grey to black in colour.

Table 4.1. (continued).

Ise	Sandstone	Ijesa-Ijebu, Abeokuta	Conglomerate, medium- to coarse-grained, cross-bedded, conglomeratic/pebbly sandstone (sometimes micaceous), with interbedded/intercalated kaolinitic claystone. Unconformably overlies the basement complex rock and generally shows a finning upward sequence.
Cretaceous Benin Flank (CBF)	Sandstone, shale, limestone	Ori-Orhin Elerin-Nla, Okeluse	Fine- to medium- to coarse-grained, sometimes conglomeratic/pebbly, cross-bedded sandstone. Well indurated at some point with large pebbles embedded in the sand. The sandstone is sometimes ferruginous. Limestone and shale are present in Okeluse

of 30 μm for optical examination. The prepared thin sections were examined with transmitted-light microscope equipped with a camera. Eleven samples were further polished and subjected to cold cathodoluminescence (CL) analysis (5 limestones) for further examination of authigenic minerals and reflected-light analyses for the identification of iron oxide minerals (6 sandstones). The CL analysis was done at an acceleration voltage of 10 – 20 kV and a current of 325 – 350 μA. Iron cemented samples were examined with a reflected-light microscope under oil immersion for easy identification and semi-quantitative distinction of hematite and goethite (Kettanah et al., 2015; Oksuz & Kocak, 2016). Carbonate mineral phases were identified by staining the slides with a mixture of 6 grams of alizarin red S and 4 grams of potassium ferricyanide in 300 ml of 0.5% hydrochloric acid (Dickson, 1978).

Grain size parameters were determined by measuring the long axis of a minimum of 100 grains on a grid adjusted to the maximum observed grain size per sample (Busch et al., 2018a). Sorting and skewness were calculated after Folk and Ward (1957) based on these measurements.

A semi-automated Pelcon point counter mounted on a transmitted-light microscope was used for point counting (300 counts), with the step length adjusted to the maximum grain size. The ternary plot of quartz (Q), feldspar (F), and rock fragments (R) was used for the classification of the sandstone samples (McBride, 1963). The limestones were classified based on the textural features and composition after Dunham (1962) and Wright (1992b). The average point count and grain size results are given as arithmetic means (supplementary material 4.2).

The intergranular volume (IGV) was calculated from the point count data as the sum of total intergranular porosity, pore-filling cements and matrix (Paxton et al., 2002). Also, the bitumen was added to the IGV calculation in the oil-sand samples. Compactional porosity loss (COPL) and cementation porosity loss (CEPL) were calculated using equation 1 and 2 after (Lundegard, 1992). Where $\emptyset_{initial}$ is the initial porosity taken as 48% (Atkins & McBride, 1992) and V_{cement} is the volume of cement calculated from point count data. The minus cement porosity (P_{mc}) is the sum of intergranular porosity and pore-filling cements, calculated from point-count data. Samples with IGV greater than 50% are not considered.

$$\text{COPL} = \emptyset_{\text{initial}} \frac{[(100-\emptyset_{\text{initial}}) \times P_{mc}]}{100\text{-IGV}} \quad \text{(Eq. 1)}$$

$$\text{CEPL} = (\emptyset_{\text{initial}}\text{-COPL}) \times \frac{V_{\text{cement}}}{P_{mc}} \quad \text{(Eq. 2)}$$

The pre-emplacement porosity in the oil sand (also called tar sand) of the Afowo Formation was calculated as the sum of intergranular porosity and pore-filling hydrocarbons.

4.5 Results

4.5.1 Sandstone Petrography

4.5.1.1 Texture

The studied sandstones in the Eastern Dahomey Basin mostly exhibit point contacts with very few long contacts (Figs. 4.5 a – h). The detrital grains (quartz, feldspar, and rock fragments) are encased mainly by detrital clay matrix and authigenic cement (Figs. 4.5 a, e).

In the Cretaceous Benin Flank (CBF), the grain size of the sandstone samples generally ranges from fine to coarse and are poorly sorted (Figs. 4.6 a – c). The medium- to coarse-grained (CBF-C, 0.34 to 0.60 mm, avg.: 0.45 mm, n = 4), fine- to medium-grained (CBF-M; 0.14 to 0.38 mm, avg.: 0.31 mm, n = 4) and the fine-grained (CBF-F; 0.15 to 0.16 mm avg.: 0.16 mm, n = 3) facies are considered to reflect the Ise, Afowo and Araromi Formations respectively (Table 4.2).

The grain size of sandstones in the Ise Formation (excluding gravels and pebbles) is predominantly medium-grained (0.25 to 0.48 mm; avg. 0.35 mm; n = 11, Table 4.2) and the samples are poorly sorted (1.33 to 1.80, avg.: 1.50, n = 11, Figs. 4.5 a, b). The pure sandstone and oil sand of the Afowo Formation generally ranges from very fine to medium sand (0.06 to 0.51 mm, avg.: 0.21 mm, n = 17 and 0.09 to 0.70, avg.: 0.22, n = 7 respectively, Fig. 4.5 c, f) with only one coarse-grained sample in the oil sand. They are also generally poorly sorted with values ranging from 1.19 to 2.18 (avg.: 1.50, n =17) and 1.22 to 1.60 (avg.: 1.36, n = 7) for the pure and the oil sand respectively. Samples from the Akinbo-Oshosun Formation have grain sizes ranging from very fine to medium sand (0.10 to 0.36 mm; avg. 0.22 mm; n = 4) with very poor to poor sorting (1.34 to 2.13, avg.: 1.63,

Table 4.2. Summary of the modal composition and sedimentological parameters of the studied sandstone samples. Akin.-Osh. = Akinbo-Oshosun, CBF = Cretaceous Benin Flank, _C = coarse, _M = medium, _F = fine.

Formation	Lithology	Class		Quartz	Feldspar	Rock fragments	Micas	Heavy minerals	FeOx grains & ooids	Detrital Matrix	Illuviated/Infiltrated clay	Glauconite		Pyrite	Goethite	Hematite	Undifferentiated FeOx cement
Afowo	Oil-sand	Min.	Detrital	49.0	0.0	0.0	0.0	0.0	0.0	5.3	0.0	0.0	Authigenic	0.0	0.0	0.0	0.0
		Max.		67.3	0.0	1.7	1.7	0.0	1.0	15.7	0.0	0.0		1.0	0.0	1.0	0.0
		Avg.		58.2	0.0	0.4	0.5	0.0	0.4	10.0	0.0	0.0		0.4	0.0	0.3	0.0
Afowo	Sandstone	Min.		42.7	0.0	0.0	0.0	0.0	0.0	0.7	0.0	0.0		0.0	0.0	0.0	0.0
		Max.		71.3	0.3	2.3	2.7	2.0	1.0	34.0	0.0	0.0		1.0	2.7	14.7	19.7
		Avg.		59.4	0.0	0.5	0.6	0.5	0.2	9.4	0.0	0.0		0.2	0.3	1.3	1.4
Ise	Sandstone	Min.		37.3	0.0	0.0	0.0	0.0	0.0	2.0	0.0	0.0		0.0	0.0	0.0	0.0
		Max.		69.3	0.3	3.3	4.0	1.0	3.0	38.0	29.7	0.0		0.3	4.0	16.7	0.0
		Avg.		55.7	0.1	0.6	0.9	0.2	0.7	13.3	6.0	0.0		0.1	0.9	3.2	0.0
Ilaro	Sandstone	Min.		40.3	0.0	0.0	0.0	0.0	0.0	0.0	0.0	0.0		0.0	0.0	0.0	0.0
		Max.		65.3	1.0	2.0	0.3	0.3	1.0	25.0	19.3	14.3		0.3	14.0	10.7	5.7
		Avg.		57.0	0.2	0.5	0.0	0.1	0.2	8.0	6.0	0.9		0.0	2.2	3.4	0.4
Akin.-Osh.	Sandstone	Min.		41.7	0.0	0.0	0.0	0.0	0.3	1.3	0.0	0.0		0.0	0.0	0.0	0.0
		Max.		53.3	0.0	2.7	0.0	0.3	1.0	47.0	6.7	0.0		0.3	5.3	24.0	3.3
		Avg.		46.7	0.0	1.3	0.0	0.1	0.5	25.4	2.0	0.0		0.2	1.3	11.2	0.9
CBF-C	Sandstone	Min.		43.3	0.0	0.0	0.0	0.0	0.0	0.3	0.0	0.0		0.0	0.0	0.0	0.0
		Max.		72.0	3.7	0.7	0.0	0.7	0.3	11.7	5.3	0.0		0.0	22.0	4.7	1.0
		Avg.		60.7	0.9	0.2	0.0	0.2	0.1	6.7	1.9	0.0		0.0	8.1	2.3	0.3
CBF-M	Sandstone	Min.		53.0	0.0	0.0	0.0	0.0	0.0	0.0	0.0	0.0		0.0	0.0	3.0	0.0
		Max.		75.7	0.3	0.7	0.3	0.0	3.3	5.7	0.0	0.0		0.0	25.3	11.3	4.7
		Avg.		64.8	0.1	0.3	0.1	0.0	0.8	2.0	0.0	0.0		0.0	15.2	6.5	1.7
CBF-F	Sandstone	Min.		58.0	0.0	0.0	0.0	0.0	0.0	1.7	0.0	0.0		0.0	0.0	0.0	0.0
		Max.		63.0	0.0	1.3	0.0	0.0	0.7	32.3	0.3	0.0		0.0	7.3	16.0	0.0
		Avg.		60.3	0.0	0.4	0.0	0.0	0.2	12.6	0.1	0.0		0.0	2.9	7.3	0.0

Table 4.2. (continued).

Formation	Lithology	Class	undifferentiated FeOx cmt	Kaolinite cement	Bitumen	Porosity	Intergranular porosity (%)	Secondary porosity (%)	Total optical porosity (%)	Pre-emplacement porosity (%)	Compaction	Intergranular volume (%)	Minus cement porosity (Pmc)	Compactional porosity loss (%)	Cementational porosity loss (%)	Sedimentology	Mean grain size (mm)	Sorting value
Afowo	Oil-sand	Min.	0.0	0.0	6.7		5.7	0.7	11.3	18.7		25.3	19.0	25.0	0.0		0.1	1.2
		Max.	0.0	0.0	23.0		20.7	13.0	21.3	30.0		43.3	30.7	35.8	0.7		0.7	1.6
		Avg.	0.0	0.0	14.5		10.6	4.8	15.4	25.1		35.3	25.4	30.1	0.2		0.2	1.4
Afowo	Sandstone	Min.	0.0	0.0	0.0		0.3	0.0	2.7	-		28.7	0.7	11.9	0.0		0.1	1.2
		Max.	19.7	0.0	0.0		37.7	5.7	39.7	-		55.3	41.0	47.7	17.3		0.5	2.2
		Avg.	1.4	0.0	0.0		24.4	1.7	26.0	-		36.8	27.2	27.7	2.3		0.2	1.5
Ise	Sandstone	Min.	0.0	0.0	0.0		2.0	0.0	5.7	-		30.3	17.0	1.3	0.0		0.2	1.3
		Max.	0.0	1.7	0.0		29.0	3.7	29.0	-		57.0	47.3	37.3	13.1		0.5	1.8
		Avg.	0.0	0.5	0.0		17.0	0.8	17.8	-		41.0	27.7	26.8	3.3		0.3	1.5
Ilaro	Sandstone	Min.	0.0	0.0	0.0		10.7	0.0	12.0	-		34.0	16.3	8.8	0.0		0.1	1.2
		Max.	5.7	0.0	0.0		31.7	1.3	32.0	-		50.0	43.0	37.8	14.3		0.6	1.8
		Avg.	0.4	0.0	0.0		20.7	0.4	21.1	-		40.8	32.8	21.8	4.8		0.4	1.4
Akin.-Osh.	Sandstone	Min.	0.0	0.0	0.0		0.3	2.3	2.7	-		43.3	4.0	10.3	1.8		0.1	1.3
		Max.	3.3	0.0	0.0		19.0	4.3	21.3	-		52.3	42.0	45.8	19.8		0.4	2.1
		Avg.	0.9	0.0	0.0		7.6	2.9	10.5	-		48.4	23.0	29.1	11.0		0.2	1.6
CBF-C	Sandstone	Min.	0.0	0.0	0.0		14.0	0.3	15.0	-		26.0	17.0	1.9	0.2		0.3	1.4
		Max.	1.0	0.0	0.0		19.3	5.3	24.7	-		47.3	47.0	37.3	27.1		0.6	1.5
		Avg.	0.3	0.0	0.0		16.7	2.1	18.8	-		35.9	29.3	24.4	9.5		0.5	1.4
CBF-M	Sandstone	Min.	0.0	0.0	0.0		2.7	0.0	3.3	-		23.3	17.7	7.7	2.5		0.1	1.2
		Max.	4.7	0.0	0.0		13.7	1.3	13.7	-		43.7	43.7	36.8	32.0		0.4	1.3
		Avg.	1.7	0.0	0.0		8.1	0.5	8.6	-		33.4	31.4	22.7	19.2		0.3	1.2
CBF-F	Sandstone	Min.	0.0	0.0	0.0		7.7	0.7	8.7	-		33.7	7.7	17.9	0.0		0.2	1.2
		Max.	0.0	0.0	0.0		19.0	3.3	20.0	-		40.0	36.7	43.7	14.2		0.2	1.4
		Avg.	0.0	0.0	0.0		14.4	1.7	16.1	-		37.3	24.8	29.1	8.0		0.2	1.3

n= 4; Fig. 4.5 e). The sandstones of the Ilaro Formation vary from medium to coarse sand (0.12 to 0.58 mm, avg.: 0.41 mm, n = 16) and are poorly sorted (1.21 to 1.75, avg.: 1.43, n = 16; Figs. 4.5 d, e, f).

4.5.1.2 Detrital constituents

Quartz (monocrystalline and polycrystalline), feldspar (plagioclase and K-feldspar) and rock fragments (mainly sedimentary) are the main detrital components in the studied samples which can be classified as quartz arenites and a few sub-arkose and litharenite (Fig. 4.7). The majority of the samples has extraordinarily low feldspar and rock fragment contents which makes the samples to cluster at the quartz (Q) apex of the ternary plot of QFR plot (Fig 4.7). Detrital feldspar grains are partially dissolved in most samples (Fig. 4.5 c). Table 4.2 presents the summary of the point count data for the studied lithologies.

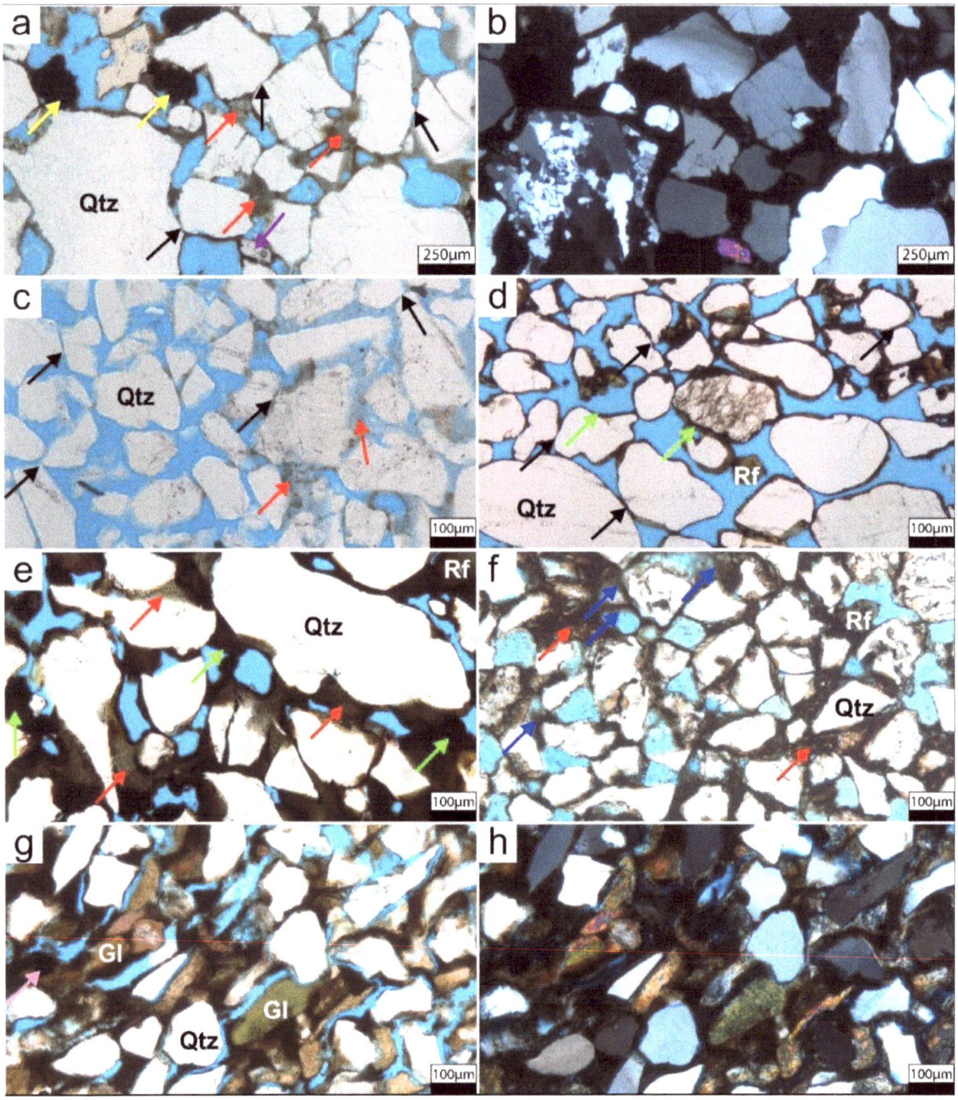

Figure 4.5. Photomicrographs of the detrital components of the siliciclastic sandstones of the Eastern Dahomey Basin. a, b) Poorly sorted, coarse-grained sandstone showing point contact (black arrows), FeOx grain (yellow arrows), detrital matrix (red arrows) and zircon (purple arrow; sample number IS_11C). c) Poorly sorted, fine-grained, weakly consolidated sandstone (black arrows show point contacts) with detrital matrix-rich layer (red arrows; sample number AF_05E). d) Poorly sorted, medium-grained sandstone with FeOx and matrix coating around the quartz (green arrow; sample number IL_A15G). e) Detrital matrix-rich (red arrows) sandstone which stabilised the framework. Green arrows show FeOx cement (sample number AK-OS_19). f) Very fine-grained poorly sorted oil sand (sample number AF_10B). Blue arrows show pore-filling oil and red arrows show oil impregnated matrix. g, h) Sandstone containing glauconite (Gl) and pyrite (pink arrow; sample number IL_A15D). Qtz = quartz, Rf = rock fragments.

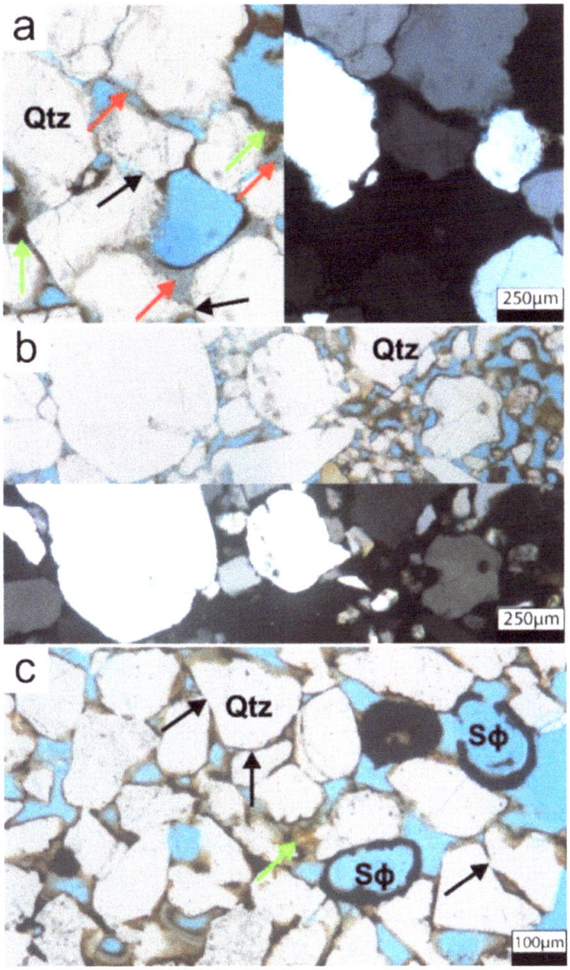

Figure 4.6. Photomicrographs of the detrital components of the siliciclastic sandstone of the Benin Flank of Eastern Dahomey Basin. a) Coarse-grained sandstone with matrix (red arrow) and FeOx stain (green arrow; CBF_20A). b) Fine- to medium-grained sandstone (CBF_22Aii). c) Fine-grained sandstone with secondary porosity (sϕ), point contact (black arrows), and brownish FeOx stain (sample number CBF_24B). Qtz = quartz.

In the Cretaceous Benin Flank (CBF), the quartz content ranges from 43.3 to 72.0% (avg.: 60.7%, n = 4), 53.0 to 75.7% (avg.: 64.8%, n =4) and 58.0 to 63.0% (avg.: 60.3%, n = 3) of the bulk sample in the coarse (CBF-C), medium (CBF-M) and fine (CBF-F) sediments respectively (Fig. 4.6) with corresponding feldspar content ranging from 0 to 3.7% (avg.: 0.9%), 0 to 0.3% (avg.: 0.1%), and 0%. Rock fragments constituting 0 to 0.3% (avg.: 0.2%), 0 to 0.3% (avg.: 0.3%) and 0 to 1% (avg.: 0.4%) are found in the CBF-C, CBF-M and CBF-F. Muscovite is not present in CBF-C and CBF-F (0%) while the value ranges from 0 to

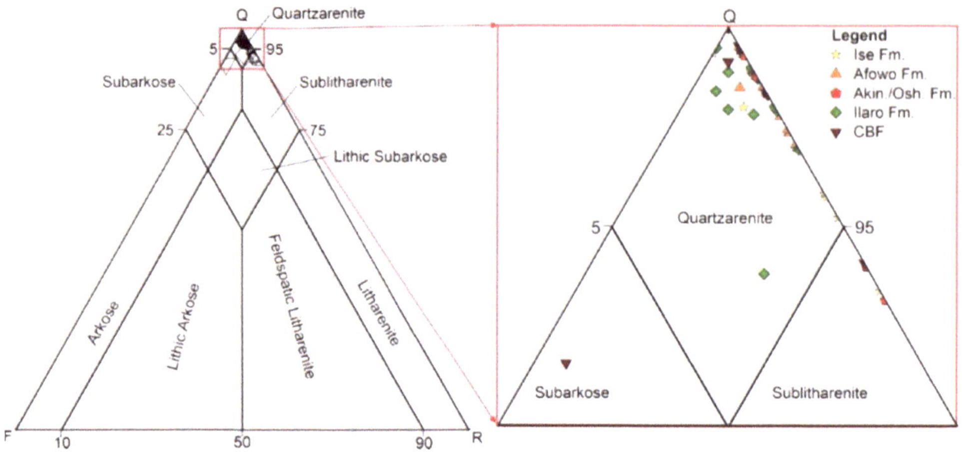

Figure 4.7. Ternary plot of quartz, feldspar and rock fragments (QFR) showing the classification of the sandstones (plotted after McBride 1963). Q = monocrystalline + polycrystalline quartz, F = Plagioclase + K-feldspar, R = all rock fragments (granitic + sandy/silty + lithic), CBF = Cretaceous Benin Flank, Fm. = Formation, Akin.-Osh. = Akinbo-Oshosun.

0.3% (avg.: 0.1%) in CBF-M. Iron oxide (FeOx) and iron oxyhydroxide (FeO(OH) grains occur in some samples but are difficult to distinguish from most ooids and are thus reported in the same class (Fig. 4.5 a, Table 4.2). Their abundance ranges from 0 to 0.3% (avg.: 0.08%, n = 4), 0 to 3.3% (avg.: 0.8, n =4) and 0 to 0.7% (avg.: 0.2, n = 3) in the CBF-C, CBF-M and CBF-F respectively (Fig. 4.8 e). The percentages of detrital clay mineral matrix ranges from 0.3 to 11.7% (avg.: 6.7%), 0 to 5.7% (avg.: 2.0%) and 1.7 to 32.3% (avg.: 12.6%) respectively in CBF-C, CBF-M and CBF-F (Figs. 4.6 a, c). Illuviated clay which sometimes forms cutans on detrital grains varies from 0 to 5.3% (avg.: 1.9%) and 0 to 0.3% (avg.: 0.1%) in CBF-C and CBF-F respectively. This occurs in different shades of brown and often shows micro-laminations (Fig. 4.8 f).

Quartz is the dominant detrital mineral in the studied sandstone samples of the Ise Formation and ranges in abundance from 37.3 to 69.3% (avg.: 55.7%, n = 11) of the total framework (Figs. 4.8 a, b). The feldspar content ranges from 0 to 0.3% (avg.: 0.1%) while rock fragments range from 0 to 3.3% (avg. 0.6%) and mica varies between 0 and 4.0% (avg.: 0.9%). The FeOx and FeO(OH) grain content ranges from 0.0 to 3.0% (avg.: 0.7%). The detrital clay mineral matrix content ranges from 2.0 to 38.0% (avg.: 13.3%) and it is predominantly kaolinitic with occasional iron oxide (FeOx) and iron oxyhydroxide (FeO(OH)) staining. The illuviated clay present in the Ise Formation ranges from 0 to 29.7% (avg.: 6.0%). The illuviated clay is also encasing all present detrital phases and is present at grain contacts (Fig. 4.8 a).

The pure sandstone of the Afowo Formation has a quartz grain content ranging from 42.7 to 71.3 (avg.: 59.4%, n = 17), feldspar content ranging from 0 to 0.3% (avg.: 0.02%) and rock fragment content ranging from 0 to 1.7% (avg.: 0.5%). The muscovite ranges from 0 to 2.7% (avg.: 0.6%) while detrital clay mineral matrix ranges from 0.7 to 34.0% (avg.: 9.4%). The content of FeOx and FeO(OH) grains ranges from 0 to 1.0% (avg.: 0.2%, n = 17) in the pure sandstone. The matrix is mainly composed of a mixture of fine silt-sized grains and clay minerals (mainly kaolinitic). The oil sand of the Afowo Formation is also mainly composed of quartz, feldspar and rock fragments ranging from 49.0 to 67.3% (avg.: 58.2%), 0% and 0 to 1.0% (avg.: 0.4%, n = 7), respectively. Muscovite ranges from 0 to 1.6% (avg.: 0.5%) and the detrital clay mineral matrix content of the oil sand ranges from 5.3 to 15.7% (avg.: 10.0%). The FeOx and FeO(OH) grains ranges from 0.0 to 1.0% (avg.: 0.4%) in the oil sand.

Sandstones of the Akinbo-Oshosun Formation consist of 41.7 to 53.3% quartz grains (avg.: 46.7%, n = 4) and 0 to 2.3% rock fragments (avg.: 1.3%). No feldspars were counted in the sandstones. The FeOx and FeO(OH) grains ranges from 0.3 to 1.0% (avg.: 0.5%, n = 4) while the detrital clay mineral matrix and illuviated clay content ranges from 1.3 to 47.0% (avg.: 25.4%) and 0 to 6.7% (avg.: 2.0%), respectively.

The quartz grain content in the Ilaro Formation ranges from 40.3 to 65.3% (avg.: 57.0%, n = 16), while feldspar ranges from 0 to 1.0% (avg.: 0.2%). The rock fragment contents range from 0 to 1.3 (avg.: 0.5%) while the muscovite content ranges from 0 to 0.3% (avg.: 0.02%). FeOx and FeO(OH) grain content ranges from 0 to 1.0% (avg.: 0.2%, n = 16). Detrital clay mineral matrix contents range from 0 to 25.0% (avg.: 8.0%) and illuviated clay contents showing micro-laminations (Figs. 4.8 c, d) ranging from 0 to 19.3% (avg.: 6.0%) are recorded in the Ilaro Formation. Glauconite grains occur in only one sample of the Ilaro Formation (14.3%; Figs. 4.5 g, h; sample number IL_A15D) where they are encased by matrix and iron oxide (FeOx) cement.

4.5.1.3 Authigenic phases in sandstones

4.5.1.3.1 Clay minerals

Kaolinite is the only authigenic clay mineral encountered and its content ranges from 0 to 1.7% (avg.: 0.5%, n = 11) in the sandstone of the Ise Formation and only in trace amount in the Cretaceous Benin Flank (CBF; Figs. 4.8 a, b, Table 4.2). It occurs as vermiform and fine-grained booklet aggregates filling intergranular pores or replacing feldspar grains.

4.5.1.3.2 Iron oxide and iron oxyhydroxide

Iron oxides (FeOx) and iron oxyhydroxides FeO(OH) are present as a non-easily distinguishable mixture of hematite and goethite (Figs. 4.8 e – h). A qualitative distinction is possible using oil immersion reflected light microscopy which indicates the presence of goethite (yellowish) and hematite (red internal reflections) in the iron oxide assemblage (Figs. 4.9 a – c).

In the Cretaceous Benin Flank (CBF), the content of FeOx and FeO(OH) ranges from 0.3 to 27.7% (avg.: 10.7%, n = 4), 4.0 to 34.7% (avg.: 23.3%, n = 4) and 0 to 17.3% (avg.: 10.2%, n = 3) respectively in CBF-C, CBF-M and CBF-F respectively.

The FeOx and FeO(OH) content of Ise Formation ranges from 0 to 16.7% (avg.: 4.2%, n = 11). In the pure sandstone and oil sand of Afowo Formation, FeOx varies from 0 to 19.7% (avg.: 3.0%, n = 17) and 0 to 1.0% (avg.: 0.3%, n = 7), respectively. The Akinbo-Oshosun Formation has FeOx content ranging from 3.3 to 24.0% (avg.: 13.4%, n = 4) while the overlying Ilaro Formation contains 0 to 16.0% (avg.: 6.0%, n = 16) FeOx. The hematite often appears as individual pigments around detrital grains (including hematite grains) and sometimes occludes pores. The goethite on the other hand mostly encases hematite and all other detrital and authigenic phases present (Fig. 4.9 c). Furthermore, hematite and goethite are sometimes found to fill inherited fractures in quartz.

4.5.1.3.3 Other authigenic components

The pore-filling solid bitumen content in the oil sand of the Afowo Formation where the oil fills the pores and impregnates the matrix ranges from 6.7 to 17.3%. Pyrite constitutes less than 1.0% in all the studied samples and it is present as framboidal crystal assemblages embedded in the detrital clay mineral matrix (Fig. 4.5 g).

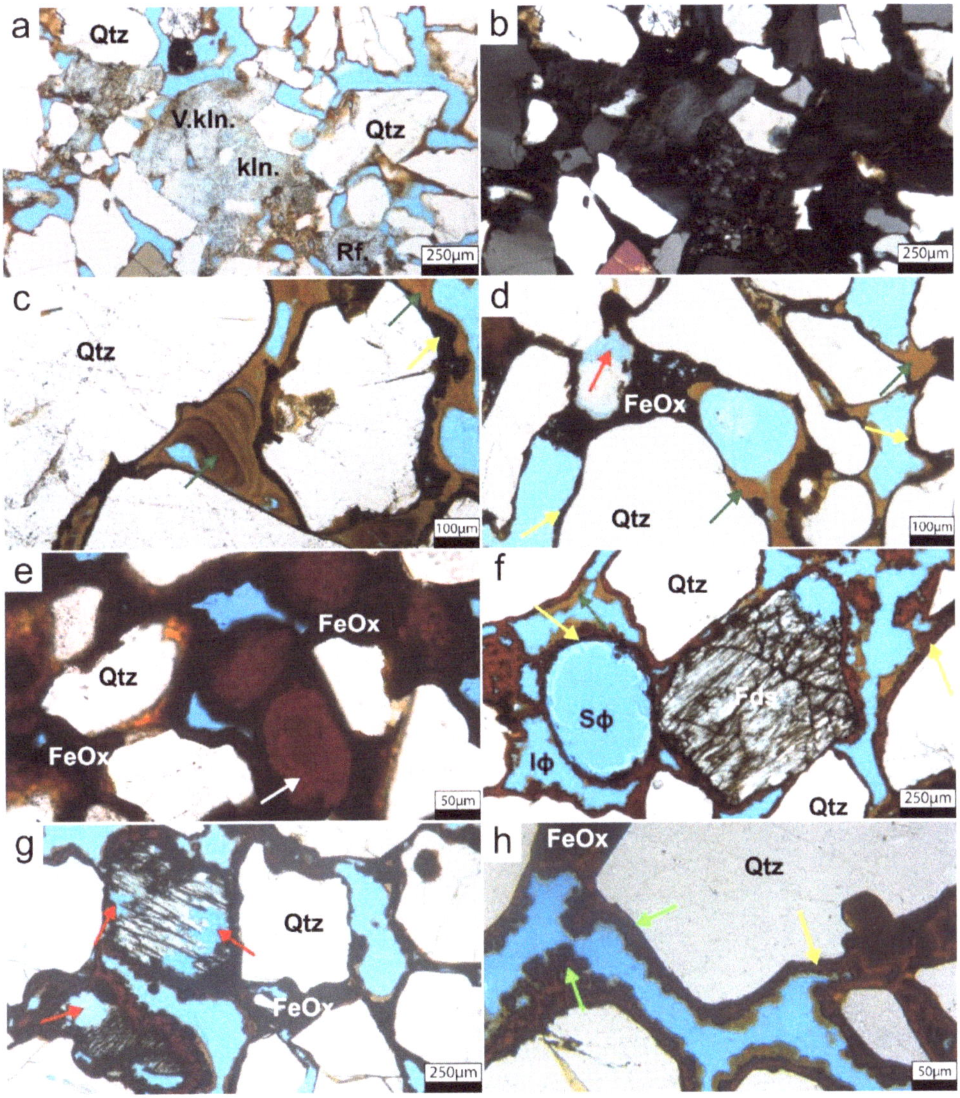

Figure 4.8. Authigenic minerals in the studied samples. a, b) Medium-grained sandstone with vermiform kaolinite and fine-grained kaolinite booklets (sample number IS_11H). c, d), FeOx coating/rim (yellow arrows), illuviated clay showing micro-lamination (green arrows), and grain dissolution which resulted in secondary porosity (red arrow) in the sandstone (sample number IL_O15B; IL_O15A). e) Iron cemented sandstone showing pore-filling and pore-lining FeOx cement with FeOx ooid (white arrow; sample number CBF_22B). f) Intragranular (secondary pore in feldspar grain and mouldic pore due to total dissolution of grain (sample number CBF_21). g) FeOx cement and secondary pore in feldspar grain (red arrows; sample number CBF_21). h) Hematite coating around detrital grains (yellow arrow) and goethite cement (green arrow; sample number CBF_21).

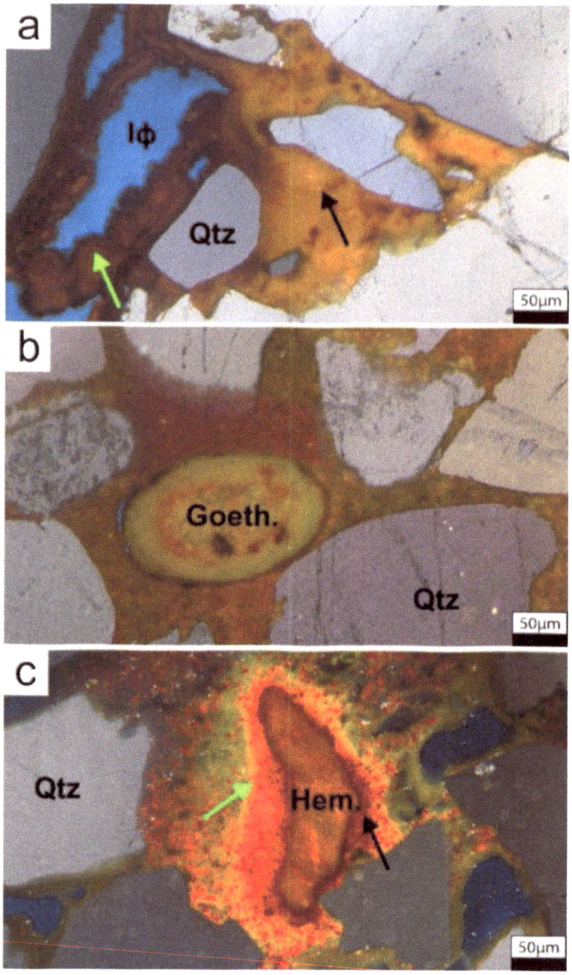

Figure 4.9. a – c) Oil immersion reflected light photomicrographs under crossed-polarized light (xpl) showing goethite (green arrow, a, c) and hematite (black arrow, a, c) cement, goethite ooid (b) and hematite grain (sample number CBF_21, c).

4.5.2 Limestone Petrography

4.5.2.1 Texture

The limestone samples are mainly composed of cement which encased the allochems/bioclasts with very few or no contacts existing between the bioclasts (Table 4.3; Figs. 4.10 a, b, d), while occasionally dolomite rhombs are the main detrital constituent (Fig. 4.10 e). The observed textures are packstone, wackestone, cementstone (mainly cement), and sparstone (crystal > 10 µm; Figs. 4.10 a – h). No indication of intense mechanical compaction (e.g. deformed or broken fossil fragments) is found in the limestone samples, while glauconite pellets are only marginally deformed in the sparstone sample (Fig. 4.10 e).

4.5.2.2 Detrital constituents

The limestones are mainly composed of bioclasts, carbonate cement, and micrite (Table 4.3). The bioclasts range in abundance from 0 to 18.3% (avg.: 6.4%, n = 8) in the Ewekoro Formation while they range from 4.7 to 7.3% (avg.: 6.0%, n = 2) in the Araromi Formation and 9.0% in the single limestone from the Cretaceous Benin Flank (CBF). These bioclast assemblages are mainly made up of foraminifera, brachiopod, echinoid, ostracod, gastropod, and other undifferentiated bioclasts (Figs. 4.10 a – h). The sparstone consists almost exclusively of dolomite rhombs and contains glauconite pellets (Fig. 4.10 e), which are indented by the detrital dolomite rhombs. Individual wackestones also contain sparry dolomite crystals embedded in sparry calcite and micrite (Fig. 4.10 f). The studied limestones also contain a minor amount of detrital quartz grains (0 to 6.0%), sedimentary rock fragments (0 to 3.0%), other undifferentiated lithoclasts (0 to 3.7%), glauconite pellets (0 – 2.0%) and 0 to 9.0% undifferentiated clay mineral matrix. The glauconite are encased in sparry calcite cement (Fig. 4.11 c) or embedded in dolomite rhombs (Fig. 4.10 e).

4.5.2.3 Authigenic composition

The main authigenic materials are sparry calcite, Fe-calcite, dolomite, Fe-dolomite and siderite occurring with 0 to 73.0%, 0 to 48.0%, 0 to 20.0%, 0 to 1.3%, and 0 to 21.0% respectively (Figs. 4.10 a – h, 4.11 a – d, Table 4.3). A few calcite veins of differing iron contents formed parallel to the bedding are also present in the Ewekoro Formation recovered from the Gbokoto and Igbeme Wells (Fig. 4.11 d). The margin of the vein contains calcite cements with less iron (red stain) than the center of the vein (purple stain, Fig. 4.11 d).

The calcite cement in the limestones is mainly poikilotopic while dolomite cement mostly encases foraminifera including their radial, bladed calcite overgrowths (Fig. 4.10 b) and siderite occurs as small crystals or overgrowing calcite (Figs. 4.10 d, f – h).

Table 4.3. Composition of the studied limestone samples. CBF = Cretaceous Benin Flank.

Formation	Class	Detrital	Bioclast	Carbonate grain	Phosphate	Detrital quartz	Other lithoclasts	Matrix	Glauconite	Authigenic	Sparry calcite	Sparry Fe-calcite	Micritic calcite	Micritic Fe-calcite	Radiaxial fibrous calcite	Calcite rim cement	Fe-calcite rim cement	Dolomite	Fe-dolomite
Araromi	Minimum		3.7	0.0	1.0	0.7	0.3	2.3	0.0		19.7	12.0	4.0	7.3	0.0	1.0	0.0	0.0	0.0
	Maximum		8.3	0.0	1.7	2.0	0.7	10.3	0.0		34.3	30.3	30.3	11.7	0.0	4.0	1.3	1.0	0.0
	Average		6.0	0.0	1.3	1.3	0.5	6.3	0.0		27.0	21.2	17.2	9.5	0.0	2.5	0.7	0.5	0.0
CBF_F			9.0	0.0	0.7	0.0	0.0	0.0	0.0		4.3	15.3	1.3	23.7	0.0	0.0	0.0	0.0	0.0
Ewekoro	Minimum		0.0	0.0	0.0	0.0	0.0	0.0	0.0		0.0	0.0	0.0	0.0	0.0	0.0	0.0	0.0	0.0
	Maximum		24.7	2.0	0.3	6.0	6.7	19.7	2.0		73.3	48.0	23.3	3.7	7.7	2.3	0.0	20.0	1.3
	Average		6.1	0.4	0.0	0.8	1.0	4.3	0.6		38.8	9.5	4.4	0.5	1.0	0.7	0.0	5.8	0.3

Formation	Class	Dolomite rhombs	Fe-dolomite rhombs	Micritic dolomite	Micritic Fe-dolomite	Siderite	Ooid	Pelloid	Pyrite	Porosity	Intergranular porosity	Secondary porosity	Total optical porosity
Araromi	Minimum	0.0	0.0	0.3	0.7	0.0	0.0	0.0	0.3		0.0	0.3	0.3
	Maximum	0.0	0.0	0.3	2.7	0.3	0.0	0.0	0.7		0.0	6.3	6.3
	Average	0.0	0.0	0.3	1.7	0.2	0.0	0.0	0.5		0.0	3.3	3.3
CBF_F		1.3	37.7	0.7	1.3	3.7	0.0	0.0	0.7		0.3	0.0	0.3
Ewekoro	Minimum	0.0	0.0	0.0	0.0	0.0	0.0	0.0	0.0		0.0	0.0	0.0
	Maximum	40.0	17.0	11.3	3.0	21.0	0.3	2.0	4.3		0.0	10.0	6.3
	Average	5.2	2.1	1.5	0.4	4.7	0.0	0.3	1.3		0.0	2.5	2.5

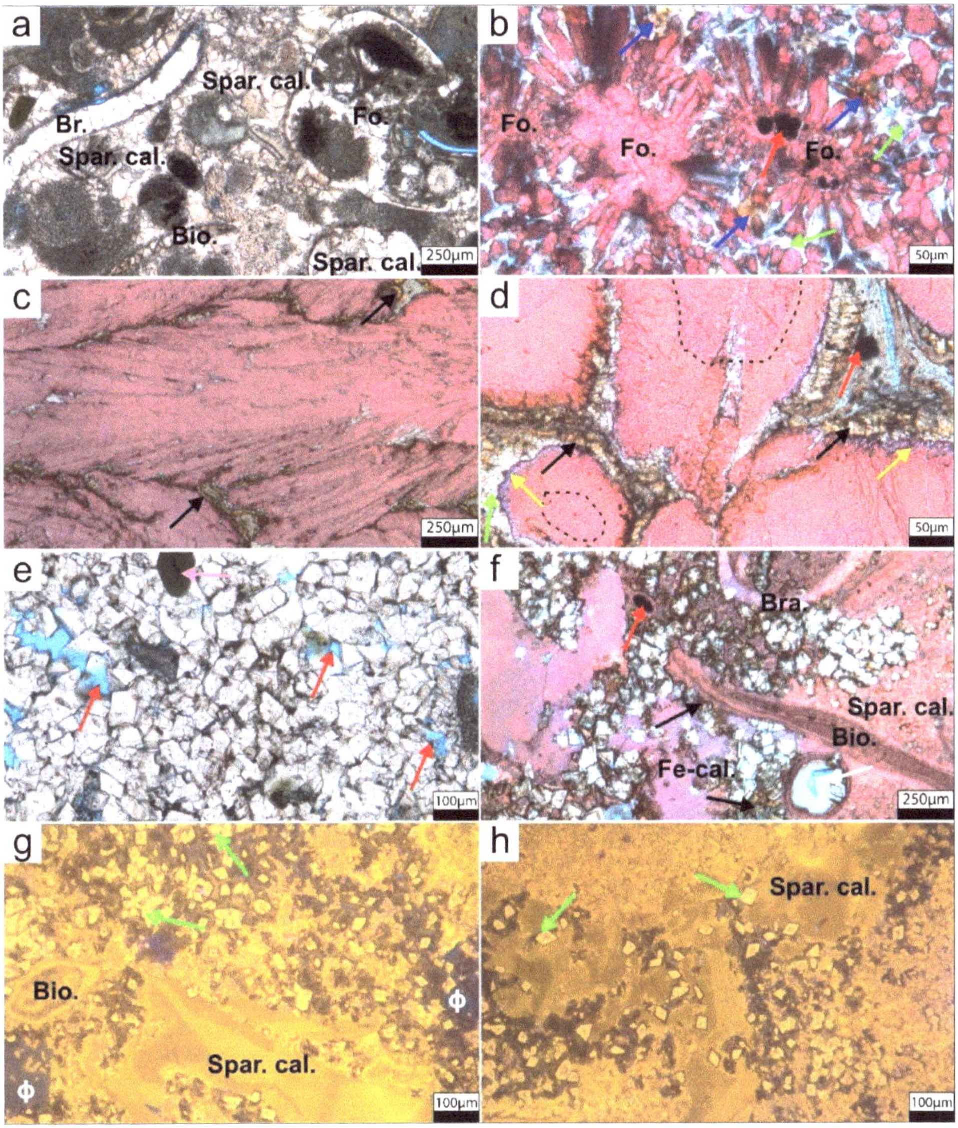

Figure 4.10. Figure caption on the next page.

Figure 4.10. Photomicrograph of the limestones. a, b) Wackestone with sparry calcite cement (stained red), dolomite cement (unstained, yellow arrows), siderite (unstained, blue arrows), fossils, and pyrite (red arrow; sample numbers EW_GB03 and EW_IG01). c) Cementstone showing fibrous calcite (stained red) with siderite rim (black arrow). d) Cementstone with Fe-rich core, Fe-rich rim (Fe-calcite; stained purple; yellow arrows), siderite rim (black arrow) and pyrite (red arrow). Green arrow show dolomite (sample number EW_GB02) while the yellow arrows shows ferroan calcite. e) Sparstone (crystalline) showing dolomite rhombs, intercrystalline porosity (red arrows) and glauconite (pink arrow; sample number EW_14B). f) Dolomitic wackestone showing distinct sparry calcite cement (stained red), dolomite rhombs (unstained), and Fe-dolomite (stained blue at the rim of dolomite rhombs), siderite (black arrow), pyrite filling foraminiferal shell (red arrow) and secondary porosity in ostracod (white arrow). g, h) Cathodoluminescence image showing sparry calcite cement, dolomite rhombs (green arrows) and pore spaces (ϕ) The darker patches are Fe-dolomite (outer edge of the rhombs) and siderite. Spar. cal. = sparry calcite, Fo = foraminifera, Bio = bioclast undifferentiated, Br. = brachiopod.

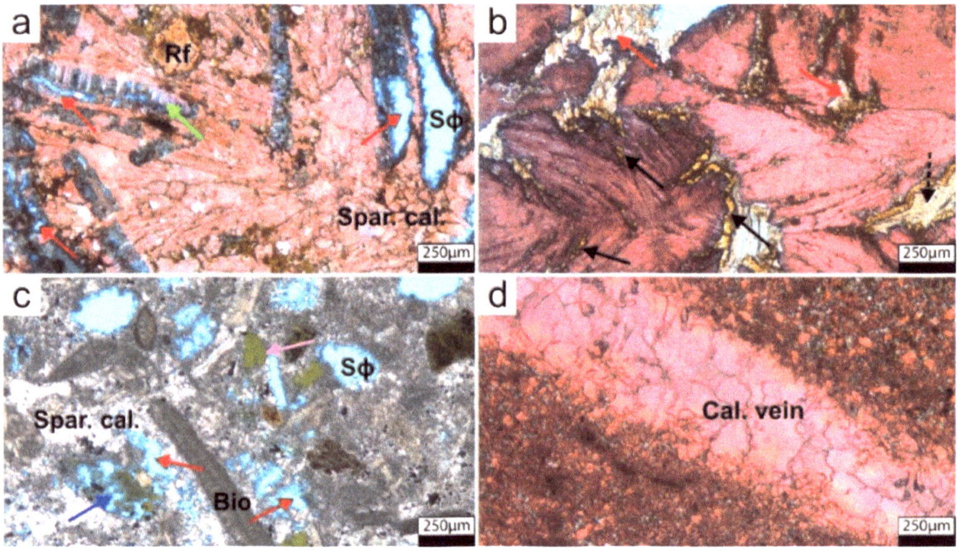

Figure 4.11. a) Secondary pores due to shell dissolution (red arrows and Sϕ). The green arrow shows Fe-calcite (ppl; sample number AR_9A). b) Photomicrograph showing carbonate paragenesis in the cementstone under plane-polarized light (ppl). The red arrow shows dolomite, the black arrow shows siderite rim growing on top of the fibrous calcite, and the black dashed arrow shows matrix (sample number IL_GB02). c) Secondary pores in limestone due to dissolution of a shell (red arrow and Sϕ, and glauconite grain (blue arrow; ppl; sample number IL_IG02). d) Ferroan calcite vein (stained puple) in limestone of calcitic composition (stained red; IG01). Sϕ = secondary pore, spar. cal. = sparry calcite, fib. cal. = fibrous calcite, cal. vein = calcite vein.

In the limestone of the Ewekoro Formation calcite sometimes forms fibrous, radial, or bladed crystals on detrital bioclasts (Fig. 4.10 b) and radiaxial fibrous aggregates (shrubs, from the Gbokoto and Igbeme wells, Figs. 4.10 c, d). The fibrous calcite in the radiaxial fibrous aggregate limestone has an inner zone composed of Fe-calcite, subsequent calcite, and an outer edge composed of ferroan calcite (Fe-calcite; Fig. 4.10 d). This is encased by dolomite and subsequently fibrous siderite rim and cement (Figs. 4.10 c, d, 4.11 b).

The fibrous, radial, or bladed crystals on detrital bioclasts are also found in the Araromi Formation but are poorly developed (Fig. 4.10 d), with a higher abundance of fossils than the Gbokoto and Igbeme wells. Another subordinate authigenic mineral in the limestone is pyrite (0 to 4.3%; Figs. 4.10 b, d; Table 4.3). Pyrite appears as framboids filling some voids (e.g. in detrital bioclasts; Figs. 4.10 a, b, f) or as cubes embedded in clay mineral matrix (Fig. 4.10 d). In the wackestone, Fe-dolomite is found encased in Fe-calcite and calcite encased in the Fe-calcite. However, where dolomite rhombs are present in the in the wackestone, the dolomite rhombs are encased by Fe-dolomite which are encased by Fe-calcite and calcite (Fig. 4.10 f). The remaining intercrystalline porosity between Fe-dolomite rhombs in this wackestone is filled by siderite. The dolomite rhombs in the wackestone sometimes have an outer edge composed of ferroan dolomite (Fe-dolomite; Fig. 4.10 f).

4.5.3 Optical porosity

4.5.3.1 Sandstones

The observed optical pore types in the studied samples are intergranular, intragranular, and intercrystalline in kaolinite (Figs. 4.8 a – h, 4.5 e; Table 4.2). In the Cretaceous Benin Flank (CBF), the total porosity ranges from 15.0 to 24.7% (avg.: 18.8%, n = 4), 3.3 to 13.7% (avg.: 8.6%, n = 4) and 8.7 to 20.0% (avg.: 16.1%, n = 3) in the coarse (CBF-C), medium (CBF-M) and fine (CBF-F) grained sandstone respectively. The corresponding intergranular and intragranular porosity accounts for 14.0 to 19.3% (avg.: 16.7%) and 0.3 to 5.3% (avg.: 2.1%), 2.7 to 13.7% (avg.: 8.1%) and 0 to 1.3% (avg.: 0.5%), 7.7 to 19.0% (avg.: 14.4%) and 0.7 to 3.3%.

The total porosity in the Ise Formation ranges from 5.7 to 29.0% (avg.: 17.9%, n = 11) in which intergranular porosity ranges from 2.0 to 29.0% (avg.: 17.0%) while intragranular porosity ranges from 0 to 3.7% (avg.: 0.8%, n = 11). In the pure sandstone of the Afowo Formation, total porosity ranges from 2.7 to 39.7% (avg.: 26.0%, n = 17) out of which intergranular and intragranular porosity accounts for 0.3 to 37.7% (avg.: 24.4%) and 0 to 5.7% (avg.: 1.7%), respectively. The oil sand of the Afowo Formation has total porosity between 11.3 and 21.3 (avg.: 15.4%, n = 7) with intergranular and intragranular porosity ranging from 5.7 to 20.7% (avg.: 10.6) and 0.7 to 13.0% (avg.: 4.7%), respectively. However, the pre-emplacement porosity in the oil sand of the Afowo Formation ranges from 18.7 to 30.0% (avg.: 25.1%, n = 7). Porosity ranging between 2.7 and 21.3% (avg.: 10.5%, n = 4) are recorded in the Akinbo-Oshosun Formation, with intergranular and intragranular porosity ranging from 0.3 to 19.0% (avg.: 7.6%) and 2.3 to 4.3% (avg.: 2.9%) respectively. Of the 12.0 to 32.0% (avg.: 21.1%, n = 16) porosity in the Ilaro Formation, intergranular porosity accounts for 10.7 to 31.7% (avg.: 20.7%) while intragranular porosity ranges from 0 to 1.3% (avg.: 0.4%).

4.5.3.2 Limestones

Intraparticle and intracystalline porosity resulting from shell and cement dissolution is the main porosity type observed in the limestone samples (Figs. 4.11 a – c). The total porosity in the limestones ranges from 0.3 to 6.3% (avg.: 2.4%, n = 2) in the Araromi Formation, 0 to 6.3% (avg.: 2.5%, n = 8) in the Ewekoro Formation and the single sample from the CBF-F has 0.3% porosity (Table 4.3).

4.5.4 Mechanical compaction, intergranular volume and porosity loss

The intergranular volume (IGV) in the sandstone samples ranges from 30.3– 57.0% (avg.: 41.0%, n = 11), 28.7 to 55.3% (avg.: 36.8%, n = 17) and 25.3 to 43.3% (avg.: 35.3%, n = 7) in the Ise Formation, pure sandstone, and oil sand of the Afowo Formation respectively (Table 4.2). Akinbo-Oshosun and Ilaro Formations' IGV ranges from 43.3 to 52.3% (avg.: 48.4%, n = 4) and 34.0 to 50.0% (avg.: 40.8%, n = 16) respectively. The Cretaceous Benin Flank (CBF) has IGV values of 26.0 to 47.3% (avg.: 35.9%, n = 4), 23.3 to 43.7% (avg.: 33.4.4%, n = 4) and 33.7 to 40.0% (avg.: 37.3%, n = 3) in coarse (CBF-C), medium (CBF-M) and fine (CBF-F) grained sandstone, respectively. The compactional porosity loss (COPL) ranges from 1.3 to 37.4% (avg.: 26.8%, n = 8) in Ise Formation, 25.0 to 35.8% (avg.: 30.1%, n = 7) in oil-sand of the Afowo Formation, 11.9 to 47.7% (avg.: 27.7%, n = 15) in pure sandstones of the Afowo Formation. The value ranges from 10.3 to 45.8% (avg.: 29.1%, n = 2) and 8.8 to 37.9% (avg.: 21.8%, n = 15) in the Akinbo-Oshosun and Ilaro Formations respectively. In the CBF, the value ranges from 1.9 to 37.4% (avg.: 24.4%, n = 3), 7.7 to 36.8% (avg.: 22.7%, n = 4) and 17.9 to 43.7% (avg.: 29.1%, n = 3) respectively in coarse (CBF-C), medium (CBF-M) and fine (CBF-F) respectively. On the other hand, cementation porosity loss (CEPL) ranges from 0 to 13.0% (avg.: 3.3%, n = 8), 0 to 0.7% (avg.: 0.2%, n = 7), and 0 to 17.3% (avg.: 2.3%, n = 15) in the Ise Formation, oil sand, and pure sandstone of the Afowo Formation respectively. The Akinbo-Oshosun and Ilaro Formations have CEPL of 1.8 to 19.8% (avg.: 11.0%, n = 2) and 0 to 14.3% (avg.: 4.8%, n = 15) respectively. The coarse, medium and fine sandstone of the CBF recorded CEPL of 0.2 to 32.0% (avg.: 9.5%, n = 3), 2.5 to 32.0% (avg.: 19.3%, n = 4) and 0 to 14.2% (avg.: 8.1%, n = 3), respectively.

4.6 Discussion

4.6.1 Porosity distribution and relation to detrital or authigenic composition

Porosity distribution in the studied samples is not related to sorting, grain size or other sedimentological parameters (Figs. 4.12 a, b). The plot of intergranular volume against optical porosity also shows no consistent correlation (Fig. 4.12 c). However, with the colour code for the sum of goethite, hematite, and illuviated clay (GHI), two groups of samples are observed (with some outliers). These are the classes with GHI contents < 12.38% GHI (A) and >12.38% GHI (B). Similarly, the colour-coded plot for the matrix content distinguishes two groups of samples; the relatively low matrix-rich (A) and the relatively high matrix-rich (B) samples (Fig. 4.12 d). The low porosity observed in some of the low GHI samples (particularly the outliers in B of Fig. 4.12 c) can be attributed to the high detrital clay mineral matrix content (Figs. 4.12 d). Generally, the intergranular porosity is reduced with increasing GHI contents at high IGV (Fig. 4.12 c). The highest optical porosity is observed in samples of the Afowo Formation where 50% of the samples have a porosity between 15.5 to 32% (Fig. 4.12 e), while the Akinbo-Oshosun Formation has the lowest range of porosity (between 5 to 16%; Fig. 4.12).

4.6.2 Clay mineral infiltration and matrix

The infiltrated clay minerals, including laminated textures, which are also present at grain contacts (Figs. 4.5 a, 4.8 a, c, 4.8 c, d, f) and also form clay cutans in the samples are the manifestation of the mechanical infiltration process, prior to mechanical compaction. The introduction of clay and mud forms part of the diagenetic evolution of continental reservoir sandstones (Molenaar & Felder, 2018; Pittman et al., 1992; Wright, 1992) and estuarine sediments (Virolle et al., 2021). This infiltration is part of pedogenic processes which are due to the introduction of oxygenated meteoric water (rich in clay flocculates, iron compounds or organic complexes) into the sediments and can be related to paleo-surface processes (Bezerra et al., 2015; Molenaar & Felder, 2018). The meniscus features indicate that the clay flocculates settle as the downward-moving water became stagnant (Figs. 4.9 c, d). The abundance of mechanically infiltrated clay cutans can play a role in the inhibition of quartz cement in more deeply buried lithologies, which might have implications for more deeply buried samples in the region. However, available data from shallow wells (maximum ~200 m) described by Adamolekun et al. (2022) also do not show any quartz cement.

Detrital clay mineral matrix (Figs. 4.5 a, c, e, 4.6 a) also occurs in the studied samples. The detrital clay mineral matrix inhibits mechanical compaction, by maintaining high IGV, and cementation by blocking the intergranular pore space, but reduces porosity. These sandstones exhibit loose packing and the detrital clay mineral matrix constitutes the major binding material. The silty to clayey matrix material is present, as winnowing under low energy conditions is not effective. This is partly controlled by the grain size deposited in the respective depositional sub-environments (e.g. flood plain or channel deposits) (Allen, 2017; Lindholm, 1987).

4.6.3 Sandstone diagenesis

The paragenetic sequence of the Eastern Dahomey Basin is constructed for the sandstones based on petrographic evidence and mineralogical associations (Table 4.4). The prominent authigenic events in the outcrop analogues of clastic sediments in the basin are iron cementation (iron oxide; FeOx and iron oxyhydroxide FeO(OH)), kaolinite formation, feldspar dissolution, kaolinite replacing feldspar, and weak compaction.

Table 4.4. The paragenetic events of the studied sandstone samples from the Eastern Dahomey Basin and their relative timing.

Relative timing	Early	Mid	Late
Events	Kaolinite Illuviation Pyrite	Oil emplacement Hematite Goethite	Illuviation Dissolution Compaction

4.6.3.1 Clay minerals

Kaolinite is the only authigenic clay mineral encountered in the studied sediments. It was found as a pore-filling cement, feldspar replacement and in vermicular form (Figs. 4.8 a, b). Its margin is stained by goethite and is thus formed before goethite precipitation. Kaolinite prominently forms in humid-tropic soils at low pH values and is a typical near-surface diagenetic mineral either during early or uplift diagenesis in presence of meteoric water (Reeves et al., 2006). Humid conditions and precipitation in West Africa in the Late Cretaceous led to intense weathering in the region (Oboh-Ikuenobe et al., 2005; Omietimi et al., 2022).

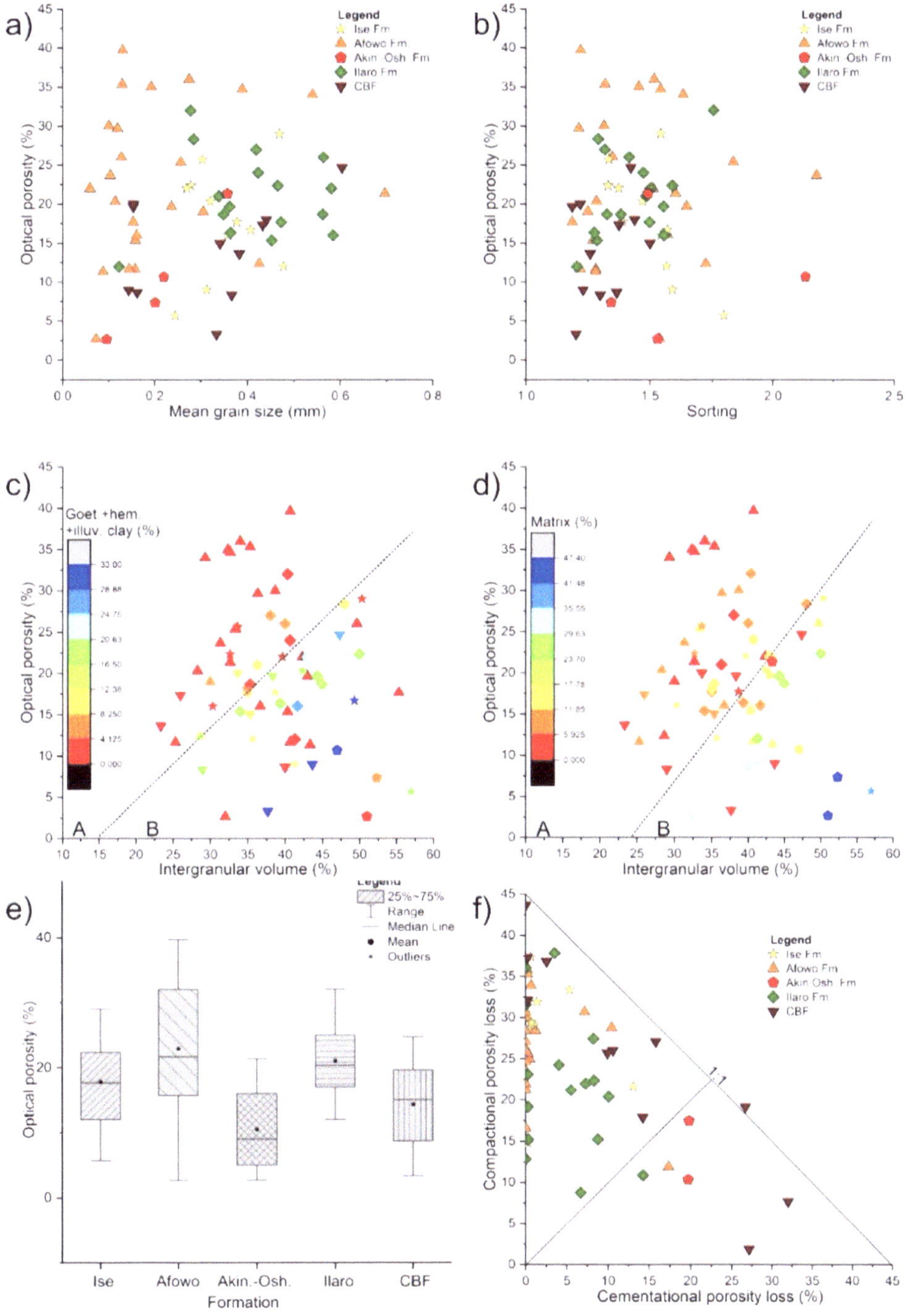

Figure 4.12. Figure caption on the next page.

Figure 4.12. a) Cross plot of optical porosity versus mean grain size showing no correlation. b) No correlation is observed between the optical porosity and sorting. c) No correlation exists in the plot of intergranular volume (IGV) versus optical porosity colour coded against the sum of goethite, hematite and illuviated clay (GHI). However, a weak positive correlation (with few outliers) is observed within the high and low GHI (A – B) where IGV increases with GHI. d) Cross plot of intergranular volume versus optical porosity, colour coded against the matrix content. e) Box chart of optical porosity within each of the studied formations showing the distribution of optical porosity in relation to the studied formation. f) Plot of compactional porosity loss (COPL) versus cementational porosity loss (CEPL) showing higher porosity loss due to compaction, while high intergranular porosities are retained (Lundegard, 1992).

4.6.3.2 Iron oxide (FeOx) and iron oxyhydroxide (FeO(OH)) phases

The formation of iron oxides is strongly controlled by the redox potential (Eh) and degree of acidity (pH). The early diagenetic hematite (iron oxide coating, Figs. 4.8 c – h) is found around quartz and other stable detrital grains. The first stage of hematite coating is succeeded by pore-lining, and pore-filling goethite (partly globular-shaped; Fig. 4.8 h) and hematite which reduced the porosity. Also, a late-stage goethite phase forms a coating on the initial hematite coating (Figs. 4.8 h). Goethite and hematite stains sometimes occur in the detrital clay/silty matrix (Figs. 4.5 d, e) and filled cleavage planes in feldspars (Figs. 4.8 f, g). Hematite and possible hematite precursors (e.g., goethite) form as grain coatings in oxidizing near-surface settings, producing reddened sediments (Scholle & Ulmer-Scholle, 2003). The source of iron could be attributed to the weathering and dissolution of iron-bearing minerals present in the Precambrian Basement Complex of Southern Nigeria (amphibole, olivine, biotite; cp. Walker et al., 1976). As FeOx phases are also present at grain contacts (e.g. Fig. 4.8 d), and previous studies often show mixed clay and FeOx (Hasner, 2004; Kämpf & Schwertmann, 1983), they may possibly also be associated to the illuviation of clay into the sediment.

The source of iron (Fe) may also be related to the dissolution of iron-rich carbonate cement, prominently observed in other samples from the Dahomey Basin (Adamolekun et al., 2022) and possible illuviation of FeOx flocculates in a setting of mixing freshwater and saline water (nearshore environment) (Kettanah et al., 2015; Scholle & Ulmer-Scholle, 2003). The formation of goethite is often related to oxidizing, humid climatic conditions, which prevailed in the studied region since the Late Cretaceous to Paleogene and at present day (Burgess et al., 2016; Maley, 2001; Oboh-Ikuenobe et al., 2005; Omietimi et al., 2022; Ramanaidou, 2009; Scholle & Ulmer-Scholle, 2003).

4.6.3.3 Pyrite

Pyrite occurs as an early and late diagenetic mineral in the glauconitic sample of the Ilaro Formation (Fig. 4.5 g). As it is present in the compacted detrital clay mineral matrix, this form is interpreted as an early diagenetic phase. The formation of pyrite is likely related to bacterial sulphate reduction (Berner, 1984) in near surface conditions.

4.6.3.4 Dissolution

Dissolution in the sandstone samples (Figs. 4.8 d, f, g) is related to K-feldspar, plagioclase, and less stable rock fragments, appearing partially or totally dissolved in the samples (Figs. 4.8 f – h). The intragranular pores and the coatings around less stable feldspar grains undergoing dissolution occasionally suggest that mechanical compaction occurs before dissolution (Figs. 4.8 f, g). As the (partially) dissolved feldspar grains still contain their clay mineral cutans and iron oxide (FeOx)/oxyhydroxides (FeO(OH) rims, dissolution postdates the infiltration of the clay minerals and formation of FeOx/FeO(OH) rims (Fig. 4.8 f, g). The dissolution pores are also devoid of clay minerals, but do contain iron oxide precipitates along cleavage plane (Fig. 4.8 f, g).

4.6.3.5 Oil emplacement

The presence of oil-stained kaolinite in the oil sand is an indication of oil migration and emplacement after kaolinite formation. The minor quantity of authigenic kaolinite in the oil sand could be a result of an early entrance of oil into the reservoir. Lambert-Aikhionbare (1982) noted an early entrance of oil into a reservoir in the Niger Delta may inhibit further clay mineral precipitation. The migration and entrapment of oil is believed to have caused a displacement of the formation water stopping clay mineral precipitation and further diagenetic alteration of the kaolinite already precipitated in the formation (De Boer et al., 1977; Lambert-Aikhionbare, 1982; Webb, 1974). Alternatively, the complete absence of feldspar grains in the studied oil sands and detection of intergranular dissolution porosity could imply that all educts of kaolinite formation have been consumed prior to oil emplacement.

4.6.3.6 Mechanical compaction

The mechanical compaction in the analysed sandstone samples generally varies from weak to fair as evident in the dominance of point contacts (Figs. 4.5 a – h) and high intergranular volume (IGV). The amount of compactional porosity loss (COPL) exceeds cementational porosity loss (CEPL; Fig. 4.12 f). Some rocks of the Afowo Formation outcrops and wells are unconsolidated to weakly consolidated. Generally, the rocks are weakly consolidated, but those exhibiting higher amounts of iron oxide content are strongly lithified. The low level of mechanical compaction results from the low overburden pressure as a result of shallow burial (present day 0 – 34.5 m). The effect of the overburden pressure is further minimized by FeOx cement, mechanically infiltrated clay minerals, and detrital clay mineral matrix (evident from high IGV; Figs. 4.5 a, h; 4.8 a, a). Adamolekun et al. (2022) reported that early diagenetic carbonate cementation contributed to weak mechanical compaction in some parts of the basin by stabilizing the framework. However, carbonate cementation appears absent in all the studied sandstone samples. This is likely due to meteoric weathering during exposure at the surface (Burgess et al., 2016).

4.6.4 Limestone diagenesis

The limestone paragenesis has been divided into the main textural classes (wackestone/packstone, cementstone, and sparstone) showing different sequences (Table 4.5). The main events in the studied limestone samples contain early diagenetic micrite formation, dolomite, Fe-dolomite, calcite, Fe-calcite, siderite, minor pyrite formation and carbonate dissolution (incl. shell fragments). There is no evidence of intense mechanical compaction in all the limestone samples.

Table 4.5. Summary of the paragenetic events of the studied limestone samples from the Eastern Dahomey Basin and their relative timing. Fe-dolomite = ferroan dolomite, Fe-calcite = ferroan calcite.

Relative timing	Early	Mid	Late
Events	Dolomite Pyrite Fe-dolomite	Fe-calcite Calcite Recrystallization	Siderite Dissolution Calcite vein

4.6.4.1 Wackestone and Packstones

Pyrite occurs as an early diagenetic mineral in the studied limestone (wackestone and packstone) where it is encased in calcite (Figs. 4.10 b, f). Shells have undergone recrystallization to non-selective isopachous calcite, ferroan calcite (Fe-calcite), and/or ferroan dolomite (Fe-dolomite; Figs. 4.10 a, b; Table 4.5). This process has led to a reduction in secondary porosity that initially resulted from shell dissolution. Dissolution in the limestones (Figs. 4.10 d, f) is mostly concentrated on fossil shells and is likely the last diagenetic modification of the limestones, as no further mineral precipitation is found in these intraparticle pores.

4.6.4.2 Sparstones

The earliest stage in the diagenetic history of the sparstone is the precipitation of Fe-dolomite on the edges of the dolomite rhombs. This is also supported by the cathodoluminescence analysis, which shows a lighter inner core and a darker outer edge of the dolomite rhombs (Figs. 4.10 g, h), indicating a relative iron enrichment in the formation water favouring precipitation of the Fe-dolomite outer edges (Xu et al., 2019; Figs. 4.10 e – h). No dolomite was found to replace calcite; thus no actual dolomitization occurred in the process.

4.6.4.3 Cementstones

The core of the radiaxial fibrous calcite aggregates show a slight iron enrichment followed by an iron-deficient middle layer and a later iron-rich outer edge (purple coloration, Fig. 4.10 d). These radiaxial fibrous calcite aggregates are typical of carbonate deposition in restricted water including lacustrine, estuarine, lagoonal, and brackish (Rodríguez-Berriguete et al., 2022; Saller et al., 2016; Wright & Baceta, 2020). Dolomite encases these radiaxial fibrous aggregates and is overgrown by siderite as the latest carbonate mineral coating the radiaxial fibrous aggregates and also encasing the dolomite rhombs (Figs. 4.10 c, d). Some fractures (parallel to bedding) are found to be filled with calcite and/or Fe-calcite cement which must have formed after the initial cementation of the cementstone (Fig. 4.11 d).

4.6.5 Porosity evolution and reservoir quality

The large range of observed optical porosity (Figs. 4.8 a – h, 4.11 a – d, Tables 4.2, 4.3) shows that different architectural elements of the same depositional system (e.g. fluvial) displayed diverse porosities. Due to the low level of mechanical compaction, high intergranular volume (Figs. 4.5 a – h, 4.8 c, d) is preserved in some samples and hence a high amount of primary porosity exists in some samples, which have low goethite, hematite, and illuviated clay mineral (GHI) and matrix contents (Figs. 4.11 c, d,). However, samples from the same formation in deeper parts of the basin (Adamolekun et al., 2022) did not show considerable amounts of goethite or hematite.

Only a minor amount of porosity was produced by dissolution contributing in part to the amount of microscopically observed pores in the sandstone samples (Figs. 4.8 d – g). This amount is relatively higher in the CBF and thus improved the porosity in the FeOx-cemented sandstone. This to a little extent compensates for the primary porosity destroyed by pore-filling FeOx. Since the amount of feldspar and less stable grains are low in all the samples 0 – 3.7% (avg.: 0.1%), further grain dissolution and/or porosity enhancement will likely be limited at greater depth.

The reservoir quality of the limestones ranges from very poor to fair based on optical examination. The poorest quality is found in the samples consisting of radiaxial fibrous aggregates found in the Gbokoto and Igbeme wells which contain fewer shells.

Authigenic phases in the studied outcrop samples differ from the subsurface samples (128 to 201 m at present-day) studied by Adamolekun et al. (2022). Iron oxide cement is largely absent, while carbonate cementation is prominent in the subsurface samples. However, samples from the same formation in deeper parts of the basin (Adamolekun et al., 2022) did not show considerable amounts of goethite or hematite, which may thus be a result of prolonged residence time in the near-surface regime with contact to meteoric water in a humid climate.

4.6.6 Exploration implications

The Eastern Dahomey Basin is heterogeneous in terms of depositional environment, lithology, sedimentological and diagenetic properties. This most likely might have been induced by changes in depositional architectural elements (e.g., floodplain vs channel infill deposits) and constant interaction of land and sea in coastal to shallow marine environments. The best reservoir quality is found in the Afowo Formation, while the worst is found in the Akinbo-Oshosun Formation. However, matrix-rich sandstone with a poor reservoir quality is also sometimes present within the Afowo Formation. The main controlling factor on reservoir quality in the studied outcrop sections are the matrix content and iron oxide (FeOx) and iron oxyhydroxide (Fe(OH)) content. The western region host better reservoir rocks compared to the eastern region of the basin. In the easternmost part of the basin, towards the Niger Delta Basin, FeOx and FeO(OH) cementation are prominent and as a result, reduced the intergranular porosity. In shallow well samples from the western part of the basin towards the Republic of Benin, carbonate contents appear to be more prominent, as early carbonate cementation stabilize the framework but reduced the intergranular porosity (Adamolekun et al., 2022). At deeper horizons, reduced porosity is expected due to increased vertical effective stress which will cause the weakly consolidated sediment to compact further.

4.7 Conclusions

The overall reservoir quality of the Cretaceous–Paleogene sandstones of the Eastern Dahomey Basin is largely affected by the clay mineral matrix content and iron oxide (FeOx) cementation. Water chemistry and sea-level changes account for the formation of iron oxide cement. The mobilization of iron in the formation water has been driven by the prevailing climatic condition since the Cretaceous and the circulation of acidic meteoric pore water, which dissolved unstable detrital grains and Fe-carbonate cements. The intense iron oxide (FeOx) and iron oxyhydroxide (FeO(OH)) cementation in some outcrop sections is related to the long exposure at the surface in a humid climate. This is because the intensity of iron cementation is expected to be reduced downhole as no pervasive iron cement was encountered in shallow exploration wells in the basin.

High detrital quartz and low matrix content favor high intergranular porosity preservation in the fluvial to shallow marine oil-bearing formation (Afowo) due to small cement contents and reduced mechanical compaction. The detrital clay mineral matrix and clay infiltration generally reduce reservoir quality. Further exploration efforts in the basin should be focused within the Afowo Formation where good porosity of 2.66 to 40% (avg.: 26%) was found, as both clay mineral matrix and goethite, hematite, and illuviated clay mineral (GHI) contents are low.

The limestones with radiaxial fibrous carbonate aggregates (cementstone) demonstrate the lowest reservoir quality, while the wackestone and sparstone have is relatively the best reservoir quality.

5 Summary

In this thesis, two basins formed on the passive margin are investigated (Fig. 5.1). The study assesses the reservoir quality in continental to shallow marine settings and identifies the possible reservoir controls.

The studied sedimentary rocks are Mesozoic to Cenozoic in age and of clastic and carbonate origin (Fig. 5.2). Sedimentation in the Dahomey Basin to the west comprises both clastic and carbonate rocks and is more heterogeneous compared to the Niger Delta in the east, which predominantly contains siliciclastic sedimentary rocks (Fig. 5.2 b). For the Niger Delta region, there are no outcrop exposures that precisely match the identified reservoirs based on seismic and well log analysis.

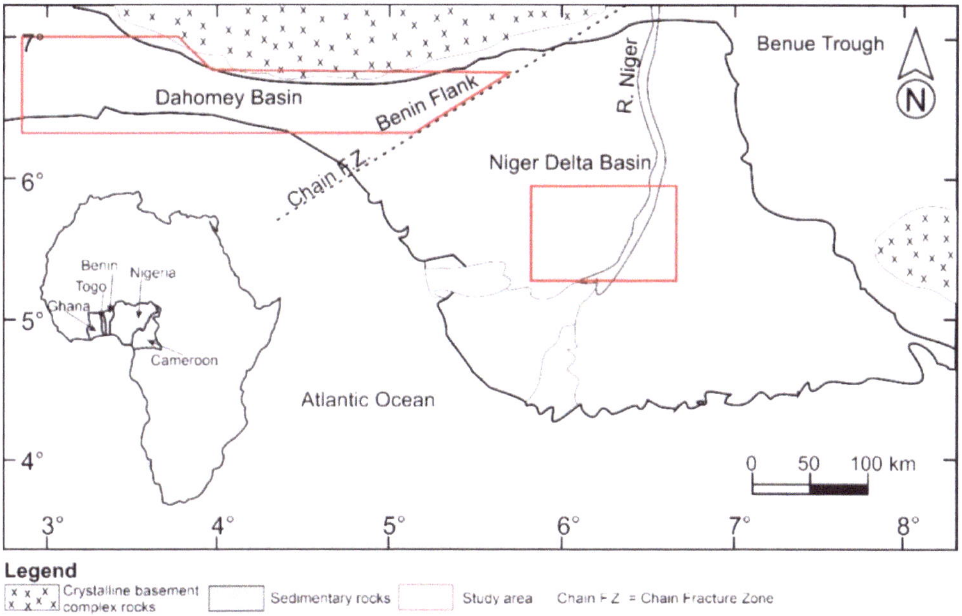

Figure 5.1. Map of southern Nigeria showing the two areas of investigation of this thesis. The diagenetic and reservoir quality prediction was done west of the Chain Fracture Zone (Dahomey Basin) while the reservoir modelling was conducted east of the Chain Fracture Zone (Niger Delta Basin). The inset map shows the location of Nigeria on the African continent.

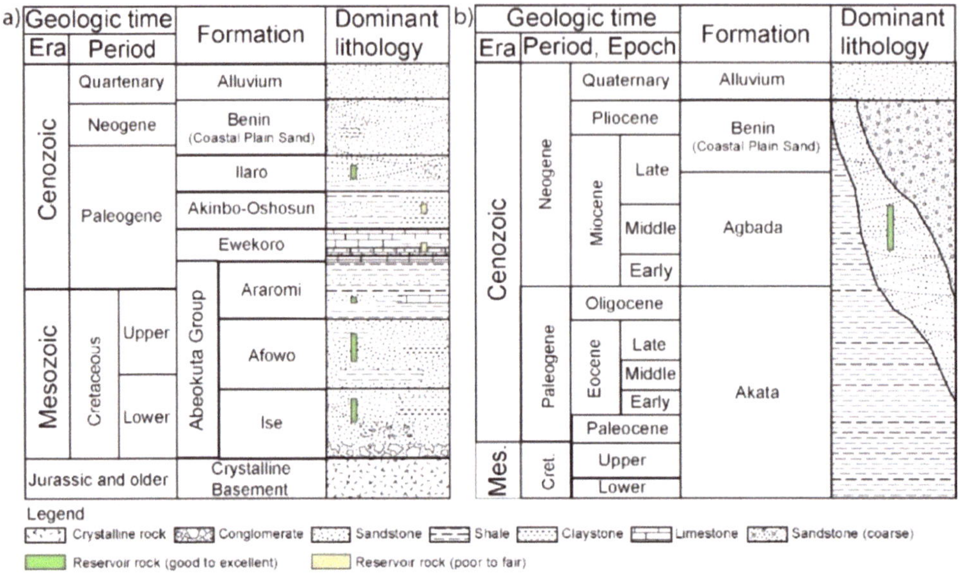

Figure 5.2. a) Stratigraphic column of Eastern Dahomey Basin. b) Stratigraphic column of the Niger Delta Basin.

The model for the Niger Delta presented in this thesis identifies its facies and petrophysical properties based on geophysical well logs. However, the methodology remains the same for core-based modelling provided the obtained properties are properly upscaled into the 3-D grid from seismic interpretation. The reservoir units (within the Miocene) in the studied field are judged to be good to excellent, with the best property occurring in the deepest, fluvial channel dominated EX 2.0 reservoir. The sensitivity and uncertainty analysis reveals the highest impact of the tortuosity factor, sorting, facies, cementation factor, and porosity on the reservoir volumetrics.

The Dahomey Basin on the other hand lacks seismic coverage, the few available lines could not be obtained from the operating companies in the region. However, based on the available shallow core samples and outcrop analogs, the sediments are largely within the early diagenetic realm and varied laterally and spatially in their authigenic mineral assemblages. The diagenetic processes of the outcrop and subsurface are not exactly correlatable in most of the clastic sediments of this basin. The carbonate rocks in the westernmost and easternmost flanks of the basin also show strong dissimilarity in the reservoir properties when compared with those in the central part of the basin. The sedimentation at the flanks likely occurred in a more restricted water based on the occurrences of radiaxial calcite aggregates (cementstones) rather than the marine environment, which was interpreted for samples from the central part. Except for few samples from the Ewekoro and Araromi Formations that classified as subarkose, the composition of all the samples are mostly quartzarenitic (Fig. 5.3).

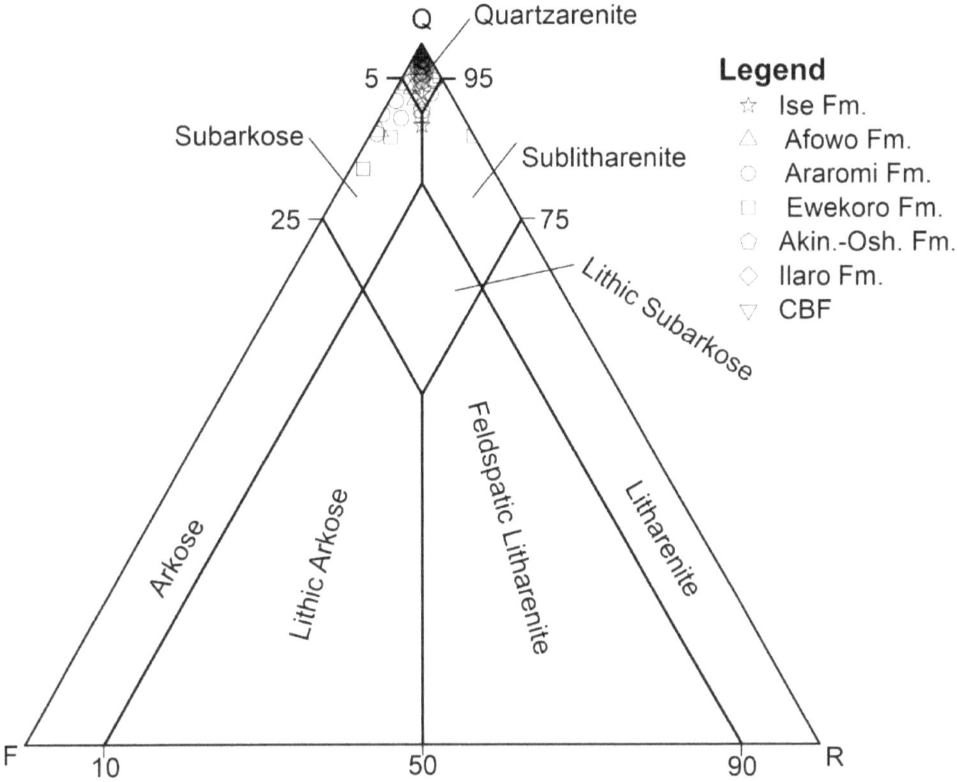

Figure 5.3. Ternary plot of quartz (Q), feldspar (F) and rock fragment (R) of the all the sandstone samples (after McBride, 1963). The samples are mainly arenitic and show no distinct mineralogical variation. Only a few samples from Ewekoro and Araromi Formations classified as subarkose.

The plot of permeability against He-porosity shows a positive correlation within the quartzarenitic, carbonate cemented and matrix-rich samples (Fig. 5.4 a). The plot of the optical porosity against intergranular volume in both the core plug and outcrop analogs shows no significant correlation (Fig. 5.4 b). The reservoir quality of this basin is judged to be good to excellent in the clastic rocks, and poor to fair in the carbonate rocks (Fig. 5.2 a).

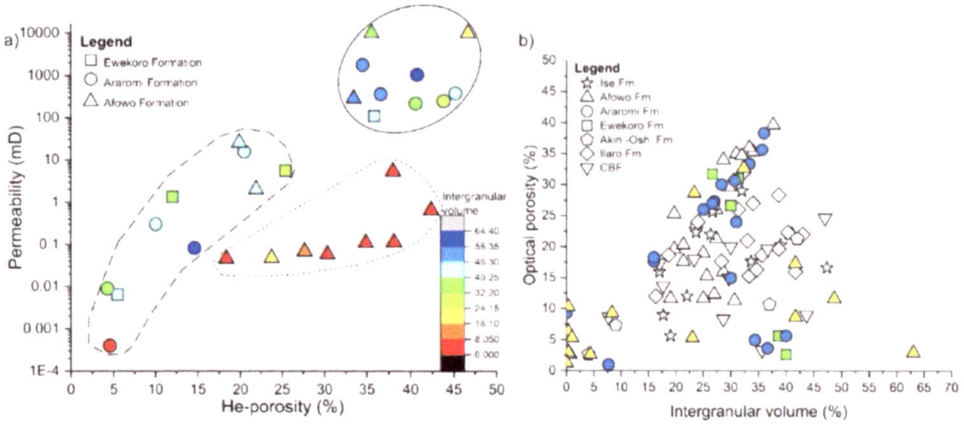

Figure 5.4. a) Permeability versus He-porosity in the core plugs showing a positive correlation. Quartz sandstone with no clay mineral matrix and no carbonate cement (solid line) has higher permeability than the remaining populations of carbonate cemented (long dash line) and clay-matrix rich (short dash) sandstones. b) Optical porosity against intergranular volume (IGV) of the core (filled symbols; green = Ewekoro, blue = Araromi, yellow = Afowo) and outcrop analog (unfilled symbols) samples.

6 Conclusions and outlook

6.1 Conclusions

The series of transgression and regression cycles account for the deposition of different packages of proximal and distal sediments in the "Atled Creek" concession in the Niger Delta Basin. The highest effective porosity is obtained in the distributary channels while the lowest values are obtained in the pro delta. Prodeltaic and delta front sediments are believed to be the most important local and regional seals in the basin while the channel and mouth bar deposits contain high reservoir quality sediments. Facies models generated by the stochastic object-based approach are important in revealing the heterogeneity of the reservoirs. Well log data and the derived facies model indicate that the reservoir units are dominated by sandstones and siltstones with subordinate shales in this deltaic setting. Normal faulting systems and the accompanying rollover anticlines, typical of extensional regimes, constitute a major structural feature of this region. The SW-area of the studied concession shows more deformation than the other parts in the field. Judging from the structural framework, good trapping mechanisms are interpreted to be present within the field. The effect of the depositional facies on petrophysical properties is accurately captured using the stochastic object-based modelling approach.

Generally, the 3-D model reveals the presence of good reservoir rocks in the studied field. Comparing the petrophysical, facies, and structural models, we conclude that unit EX 2.0 constitute the best reservoir quality in the concession with modelled effective porosities up to 28% (avg. 18.2%). The EX 2.0 reservoir unit contains oil and gas while the remaining reservoirs lying above only contain gas. Units DX 6.0, DX 7.0, DX 9.0 have modelled effective porosities of 3.3 – 24.7% (avg. 12.7%), 9.2 - 29.5% (avg. 16.7%) and 4.2 – 38.2% (avg. 13.3%), respectively. The calculated gas initially in place (GIIP) for units DX 6.0, DX 7.0, DX 9.0, and EX 2.0 respectively are: $31091*10^6$ m^3, $23682*10^6$ m^3, $22881*10^6$ m^3, $75557*10^6$ m^3, while the stock tank oil initially in place (STOIIP) in unit EX 2.0 is $214*10^6$ bbl. The identified uncertainty in the model are those associated to the tortuosity factor, sorting, facies, cementation factor, and porosity. The structural and property modelling could serve as valuable input in the initiation of a future development strategy and subsequent simulation as it unravels variability in facies, fluids, and petrophysical properties of reservoir rocks.

The reservoir quality of the siliciclastic rocks in the Eastern Dahomey Basin is largely controlled by the clay mineral matrix, carbonate, iron oxide and iron oxyhydroxide cement contents. Cementation in the basin occur as early diagenetic event during overall shallow burial. This, coupled with low overburden pressure reduced the effect of mechanical compaction, thereby maintaining mostly high IGV.

With the exception of the Ewekoro Formation, and to a lesser extent the Ilaro Formation, the matrix-rich lithofacies occurs in all formations (Ise, Afowo, Araromi, and Akinbo-Oshosun). Carbonate cementation mainly occurs as an early diagenetic event in the shallow subsurface samples of the Ewekoro Formation and parts of Araromi and Afowo Formations. While iron oxide and iron oxyhydroxide cements are present in outcrop samples from all stratigraphic intervals, the contents are higher within the rocks recovered from the outcrops in the easternmost part of the basin (Cretaceous Benin Flank). Kaolinite is the only authigenic clay mineral observed in the studied samples and occurs only in the Ise and Afowo Formations and in the Cretaceous Benin Flank.

Clay mineral matrix and mechanically infiltrated/illuviated clay fills the intergranular porosity and reduced the permeability of the rocks, while porosities are comparably high, due to preserved microporosity. The contributions of microporosity in the matrix-rich samples led to higher amount of helium porosity obtained in the core samples when compared with the corresponding optical porosity of the matrix-rich rocks. In outcrops, intense iron oxide and iron oxyhydroxide cementation occluded the intergranular porosity, thereby reducing the reservoir quality of the rocks. The mobilization of iron in the formation water is attributed to the dissolution of unstable iron-rich hinterland rocks and minerals in contact with acidic water during the marine transgression or weathering. Pervasive carbonate cementation at shallow depth also reduced the intergranular porosity of the potential reservoir rocks within the Araromi and Ewekoro Formations recovered from core material.

Intervals of good reservoir rocks are present within the Ise, Afowo, and Ilaro Formations, and the interbedded sandstone of the Araromi Formation, showing the overall best reservoir properties with porosity up to 47% and individual permeabilities above 10,000 mD.

The studied limestones in the Eastern Dahomey Basin belong to the Ewekoro and Araromi Formations, and exhibit poor to fair reservoir quality. The porosity in the limestones is restricted to dissolution of shells and carbonate cement, as well as microporosity, which is largely unconnected, resulting in low permeability (0.003 – 13 mD) at fairly high maximum porosities (7 – 32 %). Generally, the best reservoir facies in the limestone are the wackestone and packstone, while the poorest reservoir facies is the cementstone (with radiaxial calcite aggregates).

6.2 Outlook

Considering the heterogeneity in the reservoir and diagenetic properties of rocks in the continental to shallow marine environment as presented in this study, it would be interesting to examine the variation in lithologies, which have experienced deeper burial further offshore in the Niger Delta Basin. This will provide an expanded lateral insight on the nature of the sedimentological and diagenetic imprints in relation to depth in the Eastern Dahomey Basin. Also, the construction of a regional facies model is possible with the availability of deeper wells. The regional facies map would be a valuable addition in reservoir modelling and could only be reliably constructed with extensive data coverage. Also, access to deeper wells will aid the generation of more reliable burial model in the basin and in turn be important in understanding the type of clay mineral cements and the extent of carbonate cementation in deeper horizons.

Subject to the availability of seismic and well log data in the region, the transition into facies and reservoir modelling with the incorporation of the core measurements presented in this study could result in greater exploration success in the basin. Investigation of the controls on oil sand occurrence in the basin might be a key to the understanding of the reasons behind many dry wells previously drilled in the Dahomey Basin. It is suggested to establish the pattern of occurrence and the relationship of the clay mineral-rich facies of this formation in relation to the occurrences of oil-sand.

A more in-depth evaluation of the diagenetic alterations in the Dahomey Basin is necessary to more reliably establish a possible correlation with the Niger Delta Basin. Also, a more quantitative evaluation of the sedimentological and diagenetic properties of the Niger Delta Basin is needed to be incorporated in the 3-D reservoir model. The construction of a regional facies map is a good first step to achieve this. The effect of facies uncertainty identified in the reservoir model of the "Atled Creek" concession in the Niger Delta Basin would be reduced through the analysis of additional wells in the field.

7 References

Abd-Allah, A. M. A., Abdel Aal, M. H., & Ghandour, A. (2012). Structural characteristics and tectonic evolution of the northwestern margin of the Nile Delta, Egypt. *Journal of African Earth Sciences, 68*, 82-95. doi:10.1016/j.jafrearsci.2012.02.006

Abdel-Fattah, M. I., Metwalli, F. I., & Mesilhi, E. S. I. (2018). Static reservoir modeling of the Bahariya reservoirs for the oilfields development in South Umbarka area, Western Desert, Egypt. *Journal of African Earth Sciences, 138*, 1-13. doi:10.1016/j.jafrearsci.2017.11.002

Adamolekun, O. J., Benjamin, B., Oluwajana, O. A., Suess, M. P., & Hilgers, C. (2023). Diagenetic evolution of continental to shallow marine Mesozoic–Cenozoic deposits, Eastern Dahomey Basin, Nigeria: Implications for reservoir quality. Journal of Applied and Regional Geology. https://doi.org/10.1127/zdgg/2023/0378

Adamolekun, O. J., Busch, B., Suess, M. P., Molenaar, N., & Hilgers, C. (2022). Petrography and reservoir quality controls in shallow transitional marine Cretaceous-Paleogene deposits in the Dahomey Basin, Nigeria. *Journal of African Earth Sciences, 186*, 104437. doi:10.1016/j.jafrearsci.2021.104437

Adamolekun, O. J., & Umoh, E. E. (2016). Occurrence and Environmental Implications of Calcareous Nannofossils in Surface Sediments of the Western Gulf of Guinea: off Lagos Coast, South-western Nigeria. *Journal of Environment and Earth Science, 6*(6), 114-128.

Adediran, S. A., Adegoke, O. S., & Oshin, I. O. (1991). The continental sediments of the Nigerian Coastal Basins. *Journal of African Earth Sciences, 12*(1/2), 79-84.

Adegoke, O. S. (1969). Eocene Stratigraphy of Southern Nigeria. *Bulletining Bureau de Research Geologic ET Miners Memoir, 69*, 23-48.

Adeigbe, O. C., & Alo, O. O. (2017). Structural Interpretation, Trapping Styles and Hydrocarbon Potential of Block-X, Northern Depobelt, Onshore Niger Delta. *Geology, 104*, 45748-45754.

Adeigbe, O. C., & Oyekola, C. B. (2019). Geology, Palynomorphs Distribution, Stratigraphy and Depositional Environments of Lewumeji and Idogun Wells, Eastern Dahomey Basin Southwestern, Nigeria. *Journal of Geography, Environment and Earth Science International, 21*(4), 1-13. doi:10.9734/jgeesi/2019/v21i430134

Adekeye, O. A., & Akande, S. O. (2006). Depositional environments and reservoir potential assessment of the Paleogene Ewekoro Formation, eastern Dahomey Basin, southwestern Nigeria. *Journal of Mining and Geology, 42*, 133-140.

Adekeye, O. A., Akande, S. O., Bale, R. B., & Erdtmann, B. D. (2005). Carbon and oxygen isotopic compositions and diagenesis of the Ewekoro formation in the eastern Dahomey Basin, southwestern Nigeria. *Journal of Mining and Geology, 41*(1). doi:10.4314/jmg.v41i1.18832

Adelu, A. O., Aderemi, A. A., Akanji, A. O., Sanuade, O. A., Kaka, S. I., Afolabi, O., . . . Oke, R. (2019). Application of 3D static modeling for optimal reservoir characterization. *Journal of African Earth Sciences, 152*, 184-196. doi:10.1016/j.jafrearsci.2019.02.014

Adeoti, L., Onyekachi, N., Olatinsu, O., Fatoba, J., & Bello, M. (2014). Static Reservoir Modeling Using Well Log and 3-D Seismic Data in a KN Field, Offshore Niger Delta, Nigeria. *International Journal of Geosciences, 05*(01), 93-106. doi:10.4236/ijg.2014.51011

Agagu, O. K. (1985). *A Geological guide to bituminous sediments in southwestern Nigeria*. Department of Geology, University of Ibadan. Ibadan, Nigeria.

Ainsworth, R. B., Flint, S. S., & Howell, J. (2008). Predicting Coastal Depositional Style: Influence of Basin Morphology and Accommodation/Sediment Supply Regime within a Sequence Stratigraphic Framework. In G. J. Hampson, R. Steel, P. Burgess, & R. Dalrymple (Eds.), *Recent Advances in Shallow-marine Stratigraphy* (Vol. 90, pp. 237-263): Society for Sedimentary Geology, Special Publication.

Akaegbobi, I. M., Gaupp, R., & Fischer, C. (2003). *Sediment petrographic characterisation and diagenetic pathways of the Miocene Agbada sandstone, Niger delta*. Paper presented at the Annual International Conference and Exhibitions, Port Harcourt (Nigeria).

Akinmosin, A., Bankole, S. I., & Eyogwe, S. O. (2019). Reservoir geology and effects on exploitation of natural bitumen deposits (Nigerian deposits as a case study). *Journal of Petroleum Exploration and Production Technology, 9*(1), 191-205. doi:10.1007/s13202-018-0525-5

Ako, B. D., Adegoke, O. S., & Petters, S. W. (1981). Stratigraphy of the Oshosun Formation in southwestern Nigeria. *Journal of Mining and Geology, 17*(1), 97-106.

Allen, P. A. (2017). Sediment Routing Systems. The Fate of Sediments from Source to Sink. *Cambridge University Press, Cambridge*. doi:10.1017/9781316135754

Almon, W. R., & Davies, D. K. (1978). *Clay technology and well stimulation*. Paper presented at the Transsactions, Gulf Coast Assoc. Geol. Soc., USA. http://archives.datapages.com/data/gcags/data/028/028001/0001.htm

Alo, O. O., & Ehinola, O. A. (2018). Facies and Petrophysical Modelling of Sand R700, 'Sigma Field', Onshore Niger Delta: Implication on in-fill well placement *Geoscience, 114*, 1-7.

Anovitz, L. M., & Cole, D. R. (2015). Characterization and Analysis of Porosity and Pore Structures. *Reviews in Mineralogy and Geochemistry, 80*(1), 61-164. doi:10.2138/rmg.2015.80.04

Archie, G. E. (1942). The electrical resistivity log as an aid in determining some reservoir characteristics. *Journal of Petroleum Technology, 1*, 55-62.

Arribas, J., & Tortosa, A. (2003). Detrital modes in sedimenticlastic sands from low-order streams in the Iberian Range, Spain: the potential for sand generation by different sedimentary rocks. *Sedimentary Geology, 159*(3-4), 275-303. doi:10.1016/s0037-0738(02)00332-9

Asquith, G., & Krygowski, D. (2004). *Basic Well Log Analysis* (E. A. Mancini Ed.). Tulsa, Oklahoma.

Atkins, J. E., & McBride, E. F. (1992). Porosity and Packing of Holocene River, Dune, and Beach Sands. *AAPG Bulletin, 76*(3), 339-355. doi:https://doi.org/10.1306/BDFF87F4-1718-11D7-8645000102C1865D

Atlas, D. (1979). *Log Interpretation Charts.* Houston Dresser Industries, Inc.

Bahrig, B. (1994). The formation of alternating chalwsiderite units during the Neogene in the black sea - an example of climatic control in early diagenetic processes. In K. H. Wolf & G. V. Chilingarian (Eds.), *Diagenesis IV: Developments in Sedimentology* (Vol. 51, pp. 133-151). Amsterdam: Elsevier Science B.V.

Bata, T., Parnell, J., Samaila, N. K., & Still, J. (2017). Impact of oil emplacement on diagenesis in Cretaceous oil sands. *Bulletin of Canadian Petroleum Geology, 65*(2), 327-342.

Becker, I., Wüstefeld, P., Koehrer, B., Felder, M., & Hilgers, C. (2017). Porosity and permeability variations in a tight gas sandstone reservoir analogue, westphalian D, Lower Saxony Basin, NW Germany: Influence of depositional setting and diagenesis. *Journal of Petroleum Geology, 40*(4), 363-390.

Berner, R. A. (1984). Sedimentary pyrite formation: An update. *Geochimica et Cosmochimica Acta, 48*(4), 605-615. doi:10.1016/0016-7037(84)90089-9

Bezerra, C. E. E., Ferreira, T. O., Romero, R. E., Mota, J. C. A., Vieira, J. M., Duarte, L. R. S., & Cooper, M. (2015). Genesis of cohesive soil horizons from north-east Brazil: role of argilluviation and sorting of sand. *Soil Research, 53*(1). doi:10.1071/sr13188

Billman, H. G. (1976). *Offshore stratigraphy and paleontology of the Dahomey Embayment* Paper presented at the 7th African Micropaleontology Colloquium, lle-Ife.

Billman, H. G. (1992). Offshore Stratigraphy and Paleontology of the Dahomey Embayment, West African. *Nigerian Association of Petroleum Explorationist (NAPE) Bulletin, 7*(2), 121-130.

Bjørkum, P. A., & N., G. (1988). An isochemical model for formation of authigenic kaolinite, K-feldspar, and illite in sediments. *Journal of Sedimentary Petrology, 58*, 506-511.

Bjørlykke, K. (1998). Clay mineral diagenesis in sedimentary basins; a key to the prediction of rock properties; examples from the North Sea Basin. *Clay Minerals, 33*, 15-34.

Bjørlykke, K. (2015). *Petroleum Geoscience*: Springer-Verlag Berlin Heidelberg

Bjørlykke, K., & Jahren, J. (2010). Sandstones and Sandstone Reservoirs. *Petroleum Geoscience*, 113-140.

Bloch, S. (1994). Effect of Detrital Mineral Composition on Reservoir quality. . In M. D. Wilson (Ed.), *Reservoir quality assessment and prediction in clastic rocks* (Vol. 30, pp. 161-182): Society of Economic Paleontologists and Mineralogists.

Boggs Jr., S. (2009). *Petrology of Sedimentary Rocks* (Vol. Second edition): Cambridge University Press.

Boyd, G. A., Wallace, M. W., Holdgate, G. R., & Gallagher, S. J. (2004). *Marine clay and porosity evolution in the Nullawarre Greensand, Otway Basin, Southeastern Australia*. Paper presented at the PESA Eastern Australasian Basins Symposium II, Adelaide.

Boyd, R., Dalrymple, R., & Zaitlin, B. A. (1992). Classification of clastic coastal depositional environments. *Sedimentary Geology, 80*, 139-150.

Burgess, D. T., Kettler, R. M., & Loope, D. B. (2016). The Geologic Context of Wonderstone: A Complex, Outcrop-Scaled Pattern of Iron-Oxide Cement. *Journal of Sedimentary Research, 86*(5), 498-511. doi:10.2110/jsr.2016.35

Busch, B., Adelmann, D., Herrmann, R., & Hilgers, C. (2022). Controls on compactional behavior and reservoir quality in a Triassic Buntsandstein reservoir, Upper Rhine Graben, SW Germany. *Marine and Petroleum Geology, 136*. doi:10.1016/j.marpetgeo.2021.105437

Busch, B., Adelmann, D., & Hilgers, C. (2018a). *Reservoir quality modeling in deeply-buried Permian Rotliegendes sandstones, N-Germany. Impact of illite textures*. Paper presented at the EAGE Conference and Exhibition, Copenhagen.

Busch, B., Becker, I., Koehrer, B., Adelmann, D., & Hilgers, C. (2019). Porosity evolution of two Upper Carboniferous tight-gas-fluvial sandstone reservoirs: Impact of fractures and total cement volumes on reservoir quality. *Marine and Petroleum Geology, 100*, 376-390.

Busch, B., Hilgers, C., Lander, R. H., Bonnell, L. M., & Adelmann, D. (2018b). Reservoir quality and burial model evaluation by kinetic quartz and illite cementation modeling: Case study of Rotliegendes, north Germany. *AAPG Bulletin, 102*(02), 293-307. doi:10.1306/0503171605217075

Cant, D. J. (1992). Subsurface facies analysis. In R. G. Walker & N. P. James (Eds.), *Facies models, response to sea level changes* (pp. 27-45). Geol. Assoc. Canada: Canada.

Caracciolo, L. (2020). Sediment generation and sediment routing systems from a quantitative provenance analysis perspective: review, application and future development. *Earth-Science Reviews, 209*, 103226. doi:10.1016/j.earscirev.2020.103226

Coker, S. J. L. (2002). *Field Excursion Guide to Tar Sand Outcrops in Benin Basin*. Paper presented at the Nigerian Association of Petroleum Explorationists Mini-Conference.

Coker, S. J. L., Ejedawe, J. E., & Oshiorienua, J. A. (1983). Hydrocarbon source potentials of Cretaceous rocks of Okitipupa Uplift, Nigeria. *Journal of Mining and Geology, 22*, 163-169.

Curtis, C. D., & Coleman, M. L. (1986). Controls on the precipitation of early diagenetic calcite, dolomite, and siderite concretions in complex depositional sequences. *Society of Economic Paleontologist and Mineralogist, Special Publication, 38*, 23-33.

Davies, R. J., MacLeod, C. J., Morgan, R., & Briggs, S. E. (2005). Termination of a fossil continent-ocean fracture zone imaged with three-dimensional seismic data: The Chain Fracture Zone, eastern equatorial Atlantic. *Geology, Geophysics & Environment, 33*(8), 641-644.

Day, J., Clark, H., Chang, C., Hunter, R., & Norman, C. (2020). Life Cycle of Oil and Gas Fields in the Mississippi River Delta: A Review. *Water, 12*(5). doi:10.3390/w12051492

De Boer, R. B., Nagtegaal, P. J. C., & Duyvis, E. M. (1977). Pressure Solution Experiments on Quartz Sand. *Geochim. Cosmochim. Acta, 41*, 257-264.

De Matos, R. M. D. (2000). Tectonics evolution of the Equatorial South Atlantic. In W. Mohriak & M. Talwani (Eds.), *Atlantic Rifts and Continental Margins* (Vol. 115, pp. 331-354): Geophysical Monograph.

Dezfoolian, M. A., Riahi, M. A., & Kadkhodaie-Ilkhchi, A. (2013). Conversion of 3D seismic attributes to reservoir hydraulic flow units using a neural network approach: An example from the Kangan and Dalan carbonate reservoirs, the world's largest non-associated gas reservoirs, near the Persian Gulf. *Earth Sciences Research Journal, 17*(2), 75 - 85.

Dickinson, W. R., & Suczek, C. A. (1979). Plate Tectonics and Sandstone Compositions. *AAPG Bulletin, 63*, 2164-2182.

Dickson, J. A. D. (1978). Stains for carbonate minerals. In *Sedimentology* (pp. 1134-1137). Berlin, Heidelberg: Springer.

Ding, X., Han, M., & Zhang, S. (2013). The role of provenance in the diagenesis of siliciclastic reservoirs in the Upper Triassic Yanchang Formation, Ordos Basin, China. *Petroleum Science, 10*(2), 149-160. doi:10.1007/s12182-013-0262-9

Doebelin, N., & Kleeberg, R. (2015). Profex: a graphical user interface for the Rietveld refinement program BGMN. *Journal of Applied Crystallography, 48*(Pt 5), 1573-1580. doi:10.1107/S1600576715014685

Doust, H. (1990). *Petroleum geology of the Niger Delta*. London: Geological Society of London Special Publication.

Doust, H., & Omatsola, E. (1990). Niger Delta. In J. D. Edwards & P. A. Santogrossi (Eds.), *Divergent/passive Margin Basins* (Vol. 48, pp. 239-248). Tulsa: American Association of Petroleum Geologists (AAPG) Memoir.

Dunham, R. J. (1962). Classification of carbonate rocks according to depositional texture. In W. E. Ham (Ed.), *Classification of carbonate rocks* (Vol. 1, pp. 108-121): AAPG Memoir.

Dutton, S. P., Hamlin, H. S., Folk, R. L., & Clift, S. J. (1996). Early Siderite Cementation as a Control on Reservoir Quality in Submarine Fan Sandstones, Sonora Canyon Gas Play, Val Verde Basin, Texas. In L. J. Crossey, R. Loucks, & M. W. Totten (Eds.), *Siliciclastic Diagenesis and Fluid Flow: Concepts and Applications* (Vol. 55, pp. 222): SEPM Special Publication.

Egbogah, E. O., & Lambert-Aikhionbare, D. O. (1980). Possible New Potentials of the Niger delta *Oil and Gas Journal*, 176-184.

Ehinola, O. A., & Oluwajana, O. A. (2016). Conceptual basin models for Upper Senonian Araromi (marine) Formation in the Benin (Dahomey) Basin, south-western Nigeria: preliminary insight into shale play prospectivity. *Book of Abstract for Season of Omar Rahaman Students' Symposium*, 33-37.

Ekweozor, C. M. (1990). *Geochemistry of Oil Sands of SW Nigeria.* Paper presented at the Occurrence, Utilization and Economics of Nigeria tar Sands, Ago-Iwoye.

Elueze, A. A., & Nton, M. E. (2004). Organic geochemical appraisal of limestones and shales in part of eastern Dahomey Basin, southwestern Nigeria. *Journal of Mining and Geology, 40*(1), 29-40.

Enu, E. I. (1990). *Nature and Occurrence of Tar sands in Nigeria* Paper presented at the 1st workshop on Nigerian tar sands: occurrence, utilization and economic of Nigerian tar sands, Ibadan.

Estupiñan, J., Marfil, R., Delgado, A., & Permanyer, A. (2007). The Impact of carbonate cements on the reservoir quality in the Napo Fm sandstones (Cretaceous Oriente Basin, Ecuador). *Geologica Acta,, 5*(1), 89-107.

Falufosi, M. O., & Osinowo, O. O. (2021). Geology and hydrocarbon potential of the Nigerian sector of Dahomey Basin. *Journal of Sedimentary Environments*. doi:10.1007/s43217-021-00062-7

Fayose, E. A. (1970). Stratigraphy and Paleontology of Afowo-1 well, Southwestern Nigeria. *Journal Mining Geology, 5*, 1-99.

Felder, M. (2011). *Sedimentology, Petrography and Reservoir Quality of Rotliegend sandstones in Well K9ab-B5, Offshore The Netherlands* Retrieved from

Folk, R. L., & Ward, W. (1957). Brazos River bar: a study in the significance of grain size parameters. *Journal of Sedimentary Petrology, 27*, 3-26.

Garzanti, E. (2019). Petrographic classification of sand and sandstone. *Earth-Science Reviews, 192*, 545-563. doi:10.1016/j.earscirev.2018.12.014

Garzanti, E., Al-Juboury, A. I., Zoleikhaei, Y., Vermeesch, P., Jotheri, J., Akkoca, D. B., Obaid, A. K., Allen, M. B., Andó, S., Limonta, M., Padoan, M., Resentini, A., Rittner, M., Vezzoli, G. (2016). The Euphrates-Tigris-Karun river system: Provenance, recycling and dispersal of quartz-poor foreland-basin sediments in arid climate. *Earth Science Reviews, 162*, 107-128. doi:10.1016/j.earscirev.2016.09.009

George, C. F., Macdonald, D. I. M., & Spagnolo, M. (2019). Deltaic sedimentary environments in the Niger Delta, Nigeria. *Journal of African Earth Sciences, 160*. doi:10.1016/j.jafrearsci.2019.103592

Ghanbarian, B., Hunt, A. G., Ewing, R. P., & Sahimi, M. (2013). Tortuosity in Porous Media: A Critical Review. *Soil Science Society of America Journal, 77*(5), 1461-1477. doi:10.2136/sssaj2012.0435

Haque, A. K. M. E., Islam, M. A., & Ragab Shalaby, M. (2016). Structural Modeling of the Maui Gas Field, Taranaki Basin, New Zealand. *Petroleum Exploration and Development, 43*(6), 965-975. doi:10.1016/s1876-3804(16)30114-8

Harris, P. T., Heap, A., Bryce, S., Smith, R., Ryan, D., & Heggie, D. (2002). Classification of Australian clastic coastal depositional environments based upon a quantitative analysis of wave, tidal and fluvial power. *Journal of Sedimentary Research, 72*, 858-870

Hasner, K. (2004). *Untersuchungen an Hämatit-Tonmineralkrusten in Rotliegendensandsteinen des Norddeutschen Beckens,* . Jena: Friedrich Schiller Universität Jena.

Hessler, A. M., & Lowe, D. R. (2006). Weathering and sediment generation in the Archean: An integrated study of the evolution of siliciclastic sedimentary rocks of the 3.2Ga Moodies Group, Barberton Greenstone Belt, South Africa. *Precambrian Research, 151*(3-4), 185-210. doi:10.1016/j.precamres.2006.08.008

Hoffman, P., Dewey, J. F., & Burke, K. (1974). Aulacogens and their genetic relation to geosynclines with Proterozoic example from great Slave Lake, Canada. In R. H. Dott & R. H. Shaver (Eds.), *Modern and Ancient Geosynclinal sedimentation* (Vol. 19, pp. 38-55): SEPM Spec. Publ.

Houseknecht, D. W. (1988). Intergranular pressure solution in four quartzose sandstones. *Journal of Sedimentary Petrology, 58*(2), 228-246.

Huggett, J. M. (2013). Minerals: Glauconites and Green Clays. In *Reference Module in Earth Systems and Environmental Sciences*, p 1-7.

Huggett, J. M., Burley, S. D., Longstaffe, F. J., Saha, S., & Oates, M. J. (2015). The nature and origin of authigenic chlorite and related cements in Oligo–Miocene reservoir sandstones, Tapti Gas Fields, Surat Depression, offshore western India. *Journal of Petroleum Geology, 38*(4), 383-410.

Ibe, A. A., & Ezekiel, O. T. (2018). Impact of structural setting on hydrocarbon trapping mechanism of onshore Niger Delta basin, Nigeria, using seismic attribute analysis. *Egyptian Journal of Petroleum, 27*(4), 1241-1249. doi:10.1016/j.ejpe.2018.06.003

Iboyi, M., & Odedede, O. (2014). Depositional and Diagenetic Controls on Reservoir Characteristics of X-well and K-well, Ogban Field, Niger Delta. *Arab Jour. Sci. Eng., 39*, 413-422.

Ikhane, P. R., Akintola, A. I., A.O., O.-B., Oyebolu, O. O., Akintola, G. O., & Adesanwo, B. T. (2011). Palynology and paleoenvironmental reconstruction of the early Maastrichtian Afowo Formation, Dahomey Basin, Southwestern Nigeria. *Science Journal of Environmental Engineering Research, 2012*, 1-8.

Ingersoll, R. V. (1978). Petrofacies and Petrologic Evolution of the Late Cretaceous Fore-Arc Basin, Northern and Central California. . *The Journal of Geology, 86*(3), 335-352. doi:doi_10.1086_649695

Islam, M. A., Yunsi, M., Qadri, S. M. T., Shalaby, M. R., & Haque, A. K. M. E. (2020). Three-Dimensional Structural and Petrophysical Modeling for Reservoir Characterization of the Mangahewa Formation, Pohokura Gas-Condensate Field, Taranaki Basin, New Zealand. *Natural Resources Research, 30*(1), 371-394. doi:10.1007/s11053-020-09744-x

Johnsson, M. J. (1993). The system controlling the composition of clastic sediments. In M. J. Johnsson & A. Basu (Eds.), *Processes Controlling the Composition of Clastic Sediments* (Vol. 284, pp. 1-20). Colorado: Geological Society of America Special Paper

Jones, H. A., & Hockey, R. D. (1964). The Geology of part of southwestern Nigeria. *Geol. Surv. Nigeria Bull., 31*, 101.

Kämpf, N., & Schwertmann, U. (1983). Goethite and hematite in a climosequence in southern Brazil and their application in classification of kaolinitic soils. *Geoderma, 29*(1), 27-39. doi:10.1016/0016-7061(83)90028-9

Karim, A., Pe-Piper, G., & Piper, D. J. W. (2010). Controls on diagenesis of Lower Cretaceous reservoir sandstones in the western Sable Subbasin, offshore Nova Scotia. *Sedimentary Geology, 224*(1-4), 65-83. doi:10.1016/j.sedgeo.2009.12.010

Kelly, J. C., & Webb, J. A. (1999). The genesis of glaucony in the Oligo–Miocene Torquay Group, southeastern Australia: petrographic and geochemical evidence. In M. Khan, M. A. Khan, B. A. Shami, & M. Awais, 2018 (Eds.), *Microfacies analysis and diagenetic fabric of the Lockhart limestone exposed near Taxila, Margalla Hill range, Punjab, Pakistan* (Vol. 11, pp. 1-15): Arabian Journal of Geosciences.

Kettanah, Y. A., Mory, A. J., Wach, G. D., & Wingate, M. T. D. (2015). Provenance of the Ordovician–lower Silurian Tumblagooda Sandstone, Western Australia. *Australian Journal of Earth Sciences, 62*(7), 817-830. doi:10.1080/08120099.2015.1117020

Khan, Z., Sachan, H. K., Ahmad, A. H. M., Ghaznavi, A. A., & Somerville, I. (2019). Microfacies, diagenesis, and stable isotope analysis of the Jurassic Jumara Dome carbonates, Kachchh, Western India: Implications for depositional environments and reservoir quality. *Geological Journal, 55*(1), 1041-1061. doi:10.1002/gj.3470

Khani, H. F., & Back, S. (2015). The influence of differential sedimentary loading and compaction on the development of a deltaic rollover. *Marine and Petroleum Geology, 59*, 136-149. doi:10.1016/j.marpetgeo.2014.08.005

Knox, G. J., & Omatsola, E. M. (1989, 1989). *Development of the Cenozoic Niger Delta in terms of the 'Escalator Regression' model and impact on hydrocarbon distribution.* Paper presented at the KNGMG Symposium

Kou, J.-L., Tang, X.-M., Zhang, H.-Y., Lu, H.-J., Wu, F.-M., Xu, Y.-S., & Dong, Y.-S. (2012). Tortuosity for streamlines in porous media. *Chinese Physics B, 21*(4). doi:10.1088/1674-1056/21/4/044701

Lambert-Aikhionbare, D. O. (1981). *Sandstone Diagenesis and its Relation to Petroleum Generation and Migration in the Niger Delta.* (PhD). University of London, London.

Lambert-Aikhionbare, D. O. (1982). Relationship between diagenesis and pore fluid chemistry in Niger Delta oil-bearing sands. *Journal of Petroleum Geology, 4*(3), 298 1982.

Lambert-Aikhionbare, D. O., & Shaw, H. F. (1982). Significance of clays in the petroleum geology of the Niger Delta. *Clay Minerals, 17*, 91-103.

Leverett, M. C. (1941). Capillary Behavior in Porous Solids. *Transactions of the AIME, 142*(01), 152-169. doi:10.2118/941152-g

Lindholm, R. C. (1987). Chapter 8: Sedimentary rocks classification. In R. C. Lindholm (Ed.), *A Practical Approach to Sedimentology* (pp. 184-207).

Lundegard, P. D. (1992). Sandstone porosity loss - a big picture view of the importance of compaction. *Journal of Sedimebrrary Petrology, 62*(2), 250-260.

Mack, G. H. (1978). The survivability of labile light mineral grains in fluvial, aeolian and littoral marine environments: the Permian Cutler and Cedar Mesa Formations, Moab Utah *Sedimentology, 25*, 587-604.

Magbagbeola, O. A. (2005). *Sequence Stratigraphy of Niger Delta, Robertkiri Field, Onshore Nigeria.* (MSc). Texas A & M University, (Thesis_etd-tamu-2005C-GEOL-Magbagb)

Magbagbeola, O. A., & Olayinka, A. I. (2019). Sequence stratigraphy of Niger Delta - A new look at "Ake and Eta" Fields Onshore, Western Niger Delta. *Nigerian Association of Petroleum Explorationists (NAPE) Bulletin, 28*(2), 12-23.

Magbagbeola, O. A., & Willis, B. J. (2007). Sequence stratigraphy and syndepositional deformation of the Agbada Formation, Robertkiri field, Niger Delta, Nigeria. *AAPG Bulletin, 91*(7), 945-958. doi:10.1306/02050705150

Makeen, Y. M., Abdullah, W. H., Ayinla, H. A., Hakimi, M. H., & Sia, S. G. (2016). Sedimentology, diagenesis and reservoir quality of the upper Abu Gabra Formation sandstones in the Fula Sub-basin, Muglad Basin, Sudan. *Marine and Petroleum Geology, 77*, 1227-1242. doi:10.1016/j.marpetgeo.2016.08.015

Maley, J. (2001). The impact of Arid phases on the African Rain Forest through geological history. In W. Weber, L. White, A. Vedder, & L. Naughton-Treves (Eds.), *African rain forest ecology and conservation. An interdisciplinary perspective* (pp. 68-87). USA: Yale University Press.

McBride, E. F. (1963). A classification of common sandstones. *Journal of Sedimentary Petrology, 33*(3), 664-669.

McBride, E. F. (1989). Quartz Cement in Sandstones: A Review. *Earth-Science Reoiews, 26*, 69-112.

McKinley, J. M., Worden, R. H., & Ruffell, A. H. (2003). Smectite in Sandstones: A Review of the Controls on Occurrence and Behaviour During Diagenesis. *Int. Assoc. Sedimentol. Spec. Publ., 34*, 109-128. doi:10.1002/9781444304336.ch5

McLoughlin, S. (2001). The breakup history of Gondwana and its impact on pre-Cenozoic floristic provincialism. *Aust. Jour. Bot., 49*, 271-300.

Merki, P. (1970). Structural geology of the Cenozoic Niger Delta. In T. F. J. Dessauvage & A. J. Whiteman (Eds.), *African Geology* (pp. 635-646). Ibadan, Nigeria: University of Ibadan Press.

Milad, B., Ghosh, S., Slatt, R., Marfurt, K., & Fahes, M. (2020). Practical Aspects of Upscaling Geocellular Geological Models for Reservoir Fluid Flow Simulations: A Case Study in Integrating Geology, Geophysics, and Petroleum Engineering Multiscale Data from the Hunton Group. *Energies, 13*(7), 1-27. doi:10.3390/en13071604

Molenaar, N., & Felder, M. (2018). Clay Cutans and the Origin of Illite Rim Cement: An Example from the Siliciclastic Rotliegend Sandstone in the Dutch Southern Permian Basin. *Journal of Sedimentary Research, 88*(5), 641-658. doi:10.2110/jsr.2018.33

Monsees, A. C., Busch, B., & Hilgers, C. (2021). Compaction control on diagenesis and reservoir quality development in red bed sandstones: a case study of Permian Rotliegend sandstones. *International Journal of Earth Sciences, 110*(5), 1683-1711. doi:10.1007/s00531-021-02036-6

Monsees, A. C., Busch, B., Schoner, N., & Hilgers, C. (2020a). Rock typing of diagenetically induced heterogeneities - A case study from a deeply-buried clastic Rotliegend reservoir of the Northern German Basin. *Marine and Petroleum Geology, 113*, 104163. doi:10.1016/j.marpetgeo.2019.104163

Monsees, A. C., Busch, B., Schöner, N., & Hilgers, C. (2020b). Rock typing of diagenetically induced heterogeneities – A case study from a deeply-buried clastic Rotliegend reservoir of the Northern German Basin. *Marine and Petroleum Geology, 113*. doi:10.1016/j.marpetgeo.2019.104163

Morad, S. (1998). Carbonate cementation in sandstone: distribution patterns and geochemical evolution. *Int. Ass. Sediment. Spec. Publs., 26*, 1-26.

Morad, S., Al-Ramadan, K., Ketzer, J. M., & De Ros, L. F. (2010). The impact of diagenesis on the heterogeneity of sandstone reservoirs: A review of the role of depositional facies and sequence stratigraphy. *AAPG Bulletin, 94*(8), 1267-1309. doi:10.1306/04211009178

Naiman, E. (1982). Sedimentation and diagenesis of a shallow marine carbonate and siliciclastic shelf sequence: The Permian (Guadalupian) Grayburg Formation, south-eastern New Mexico. *MA, The University of Texas at Austin*, 197.

Nazeer, A., Abbasi, S. A., & Solangi, S. H. (2016). Sedimentary facies interpretation of Gamma Ray (GR) log as basic well logs in Central and Lower Indus Basin of

Pakistan. *Geodesy and Geodynamics, 7*(6), 432-443. doi:10.1016/j.geog.2016.06.006

NNPC-Report. (1997 to 2020). *Annual Statistical Bulletin*. Retrieved from http://www.nnpcgroup.com/

Nton, M. E. (2001). Aspect of rock evaluation studies of the Maastrichtian-Eocene sediments. *Journal of Mining and Geology, 13*, 33-39.

Nton, M. E., & Adamolekun, O. J. (2018). Sedimentological and Geochemical Characteristics of Outcrop Sediments of Southern Bida Basin, Central Nigeria: Implications for Provenance, Paleoenvironment and Tectonic History. *Ife Journal of Science, 18*(2), 345-369.

Nton, M. E., & Adeyemi, M. O. (2015). Petrography, compositional characteristics and stable isotope geochemistry of the Ewekoro formation from Ibese Corehole, eastern Dahomey basin, southwestern Nigeria. *Global Journal of Geological Sciences, 13*(1). doi:10.4314/gjgs.v13i1.5

Nton, M. E., Ikhane, P. R., & Tijani, M. N. (2009). Aspect of rock-eval studies of the Maastrichtian-Eocene sediments from subsurface, in the eastern Dahomey Basin southwestern Nigeria. *European Journal of Scientific Research, 25*(3), 417-427.

Nton, M. E., & Otoba, O. W. (2011). Lithofacies and organic geochemical studies of Akinside 1582 Well, Eastern Dahomey Basin Southwestern Nigeria. *NAPE Bulletin, 23*(1), 107-117.

Nton, M. E., & Salami, R. (2016). Reservoir characteristics and palaeo-depositional environment of Duski Field, onshore, Niger-Delta, Nigeria. *Global Journal of Geological Sciences, 14*(1). doi:10.4314/gjgs.v14i1.5

O'Brien, G. W., Milnes, A. R., Veeh, H. H., Heggie, D. T., Riggs, S. R., Cullen, D. J., Marschall, J. F., & Cook, P. J. (1990). Sedimentation dynamics and redox iron-cycling: controlling factors for the apatite—glauconite association on the East Australian continental margin. *Geological Society, London, Special Publications, 52*(1), 61-86. doi:10.1144/gsl.Sp.1990.052.01.06

Obaje, N. G. (2009). *Geology and Mineral Resources of Nigeria*: Springer Dordrecht Heidelberg London New York.

Oboh-Ikuenobe, F. E., Obi, C. G., & Jaramillo, C. A. (2005). Lithofacies, palynofacies, and sequence stratigraphy of Palaeogene strata in Southeastern Nigeria. *Journal of African Earth Sciences, 41*(1-2), 79-101. doi:10.1016/j.jafrearsci.2005.02.002

Oboh, F. E. (1992a). Middle Miocene palaeoenvironments of the Niger Delta. *Palaeogeography, Palaeoclimatology, Palaeoecology, 92*, 55-84.

Oboh, F. E. (1993). Depositional history of the E2.0 Reservoir in the Kolo Creek Field, Niger Delta. *Journal of Petroleum Geology, 16*(2), 197-212.

Oboh, F. E. (1995). Sedimentological and Palynological characteristics of the E2.0 Reservoir (Middle Miocene) in the Kolo Creek Field, Niger Delta. In *Geology of deltas* (pp. 243-256).

Odedede, O. (2019). Petrographic Characterization of Selected Sidewall Core Samples from NG-1 Well, Niger Delta, Nigeria: Palaeoclimatic Implication. *Journal of Scientific Research, 11*(2), 157-164. doi:10.3329/jsr.v11i2.37110

Odigi, M. I. (1994). Clay-mineral studies of the subsurface Niger Delta. *Journal of Petroleum Geology, 17*(4), 445-460.

Odin, G. S., & Matter, A. (1981). De glauconiarum origine. *Sedimentology, 28*, 611-641.

OffshoreEnergyNewsletter. (2019). First oil flows from Aje field offshore Nigeria Retrieved from https://www.offshore-energy.biz/first-oil-flows-from-aje-field-offshore-nigeria/

Ogbe, F. G. A. (1972). *Stratigraphy of Strata Exposed in the Ewekoro Quarry, Western Nigeria*. Ibadan: University of Ibadan Press

Ogbe, O. B., & Edegbai, J. A. (2022). Seismic analysis of growth faults to predict sequence stratigraphic reservoir properties: A case study of the middle to late Miocene Coastal Swamp strata, Niger Delta Basin, Nigeria. *Marine and Petroleum Geology, 138*. doi:10.1016/j.marpetgeo.2022.105536

Ogundana, A. K., Aladejana, J. A., Oladapo-Adeoye, O. O., & Talabi, A. O. (2014). Investigation of lateral continuity of bituminous sands in parts of the eastern Dahomey Basin, Nigeria. *International Journal of Current Research, 6*, 5600-5604.

Ojo, O. J. (2004). Microfacies analysis, depositional environments and diagenetic history of the limestone facies of the Pindiga Formation, Gongola Basin, Nigeria. *Jour. Min. Geol., 40*(2), 127-132.

Okoli, A. E., Agbasi, O. E., Lashin, A. A., & Sen, S. (2021). Static Reservoir Modeling of the Eocene Clastic Reservoirs in the Q-Field, Niger Delta, Nigeria. *Natural Resources Research*. doi:10.1007/s11053-020-09804-2

Okosun, E. A. (1990). A review of the Cretaceous stratigraphy of the Dahomey Embayment, West Africa. *Cretaceous Research, 11*, 11-17.

Okosun, E. A. (1998). Review of the Early Tertiary Stratigraphy of Southwestern Nigeria. *Journal of Mining and Geology, 34*, 27-35.

Oksuz, N., & Kocak, I. (2016). Geochemical evidence for the genesis of the Sarical-Yavu hematite mineralizations (Sivas, Central Turkey). *Arabian Journal of Geosciences, 9*(6). doi:10.1007/s12517-016-2432-8

Olabode, S. O. (2006). Siliciclastic slope deposits from the Cretaceous Abeokuta Group, Dahomey (Benin) Basin, southwestern Nigeria. *Journal of African Earth Sciences, 46*(3), 187-200. doi:10.1016/j.jafrearsci.2006.04.008

Olabode, S. O., & Mohammed, M. Z. (2016). Depositional Facies and Sequence Stratigraphic Study in Parts of Benin (Dahomey) Basin SW Nigeria: Implications on the Re-Interpretation of Tertiary Sedimentary Successions. *International Journal of Geosciences, 7*, 210-228. doi:10.4236/ijg.2016.72017

Oluwajana, O., Ehinola, O., Ofiwe, C., Akhayere, E., Egunjobi, K., Asanbe, J., & Akinjo, O. (2019). Microfacies Analysis and Geochemical Evaluation of Campanian-Maastrichtian Limestone Along the Benin Flank, Southwestern Nigeria. In D. M. D. e. al. (Ed.), *Petrogenesis and Exploration of the Earth's Interior* (pp. 85-87).

Oluwajana, O. A., Adebambo, B. A., Olawuyi, G. T., Ewuji, J. O., Adejayan, B. A., Ayodele, O. D., Adeniran, P. M., Arabi, D. O., & Adedokun, T. A. (2021). Paleoecologic implications of foraminiferal assemblages in the Upper Cretaceous-Paleogene strata, eastern Dahomey (Benin) Basin, southwestern Nigeria. *Arabian Journal of Geosciences, 14*(9). doi:10.1007/s12517-021-07038-x

Oluwajana, O. A., Ehinola, O. A., Ofiwe, C. U., Akhayere, E., & Egunjobi, K. (2020). Depositional environment and diagenesis of Late Cretaceous-Early Paleogene carbonates on the Benin Flank, southwestern Nigeria. *Journal of African Earth Sciences, 163*, 1-12. doi:10.1016/j.jafrearsci.2020.103762

Omatsola, M. E., & Adegoke, O. S. (1981). Tectonic evolution and Cretaceous Stratigraphy of the Dahomey Basin. *Journal of Mining and Geology, 18*(1), 130-137.

Omietimi, E. J., Lenhardt, N., Yang, R., Götz, A. E., & Bumby, A. J. (2022). Sedimentary geochemistry of Late Cretaceous-Paleocene deposits at the southwestern margin of the Anambra Basin (Nigeria): Implications for paleoenvironmental reconstructions. *Palaeogeography, Palaeoclimatology, Palaeoecology, 600*. doi:10.1016/j.palaeo.2022.111059

Onengiyeofori, A. D., Chigozie, I.-C., Dein, H. D., & Prince, S. N. (2019). Permeability Modelling of a Sandstone Reservoir in Parts of the Niger Delta. *Asian Journal of Applied Science and Technology, 3*(3), 73-89.

Onuoha, K. M., & Ofoegbu, C. O. (1988). Subsidence and evolution of Nigeria's continental margin: implications of data from Afowo-1 well. *Marine and Petroleum Geology, 5*, 175-181.

Palomares, M., & Arribas, J. (1993). Modern stream sands from compound crystalline sources: Coniposition and sand generation index. *Geological Society of America Special Paper, 280*, 313-322.

Paxton, S. T., Szabo, J. O., Ajdukiewicz, J. M., & Klimentidis, R. E. (2002). Construction of an intergranular volume compaction curve for evaluating and predicting compaction and porosity loss in rigid-grain sandstone reservoirs. *AAPG Bulletin, 86*(12), 2047–2067.

Pearce, M. A., Timms, N. E., Hough, R. M., & Cleverley, J. S. (2013). Reaction Mechanism for the Replacement of Calcite by Dolomite and Siderite: Implications for Geochemistry, Microstructure and Porosity Evolution during Hydrothermal Mineralisation. *Contributions to Mineralogy and Petrology*, 1-16.

Peralta, O. O. (2009). *Rock Types and Flow Units in Static and Dynamic Reservoir Modeling: Application to Mature Fields*. Paper presented at the SPE Latin American and Caribbean Petroleum Engineering Conference, Colombia.

Petters, S. W. (1978). Dolomitization of the Ewekoro Limestone. *Jornal of Mining and Geology, 15*(2), 78-83.

Petters, S. W. (1984). An ancient submarine canyon in the Oligocene-Miocene of the western Niger Delta. *Sedimentology, 31*, 805-810.

Pigott, J. D., & Abdel-Fattah, M. I. (2014). Seismic stratigraphy of the Messinian Nile Delta coastal plain: Recognition of the fluvial Regressive Systems Tract and its potential for hydrocarbon exploration. *Journal of African Earth Sciences, 95*, 9-21. doi:10.1016/j.jafrearsci.2014.02.003

Pittman, E. D., Larese, R. E., & Heald, M. T. (1992). Clay coats: occurrence and relevance to preservation of porosity in sandstones. In D. W. Houseknecht & E. D. Pittman (Eds.), *Origin, Diagenesis, and Petrophysics of Clay Minerals in Sandstones.* (Vol. 47, pp. 241-255): SEPM Special Publication.

Postma, D. (1981). Formation of siderite and vivianite and the pore-water composition of a recent bog sediment in Denmark. *Chemical Geology, 31*, 225-244.

Poupon, A., Clavier, C., Dumanoir, J., Gaymard, R., & Misk, A. (1970). Log Analyses of Sand-Shale Sequences- A Systematic Approach. *Journal of Petroleum Technology (JPT)*, 869-881.

Radwan, A. A., Nabawy, B. S., Abdelmaksoud, A., & Lashin, A. (2021). Integrated sedimentological and petrophysical characterization for clastic reservoirs: A case study from New Zealand. *Journal of Natural Gas Science and Engineering, 88*. doi:10.1016/j.jngse.2021.103797

Radwan, A. E. (2020). Modeling the Depositional Environment of the Sandstone Reservoir in the Middle Miocene Sidri Member, Badri Field, Gulf of Suez Basin, Egypt: Integration of Gamma-Ray Log Patterns and Petrographic Characteristics of Lithology. *Natural Resources Research, 30*(1), 431-449. doi:10.1007/s11053-020-09757-6

Ramanaidou, E. R. (2009). Genesis of lateritic iron ore from banded iron-formation in the Capanema mine (Minas Gerais, Brazil). *Australian Journal of Earth Sciences, 56*(4), 605-620. doi:10.1080/08120090902806354

Reeves, G. M., Sims, I., & Cripps, J. C. (2006). Formation and alteration of clay materials. In R. J. Pankhiurst, P. Doyle, F. J. Gregory, J. S. Oriffiths, A. J. Hartley, R. E. Holdsworth, J. A. Howe, P. T. Leat, A. C. Morton, N. S. Robins, & J. P. Turner (Eds.), *Clay Materials Used in Construction* (Vol. 21, pp. 29-71): Geological Society, London, Engineering Geology Special Publications.

Reijers, T. J. A. (1996). Structural Geology. In T. J. A. Reijers (Ed.), *Sedimentary Geology and Sequence Stratigraphy in Nigeria* (pp. 87-101). Warri, Nigeria: Shell Petroleum Development Company of Nigeria.

Reijers, T. J. A. (2011). Stratigraphy and sedimentology of the Niger Delta. *Geologos, 17*(3), 133-162. doi:doi: 10.2478/v10118-011-0008-3

Reyment, R. A. (1965). *Aspects of the geology of Nigeria - The Stratigraphy of the Cretaceous and Cenozoic Deposits*. Ibadan: Ibadan University Press.

Ringrose, P., & Bentley, M. (2015). *Reservoir Model Design*: Springer Science+Business Media B.V.

Rodríguez-Berriguete, Á., Dal' Bo, P. F., Valle, B., & Borghi, L. (2022). When distinction matters: Carbonate shrubs from the Aptian Barra Velha Formation of Brazilian's Pre-salt. *Sedimentary Geology, 440*, 106236. doi:10.1016/j.sedgeo.2022.106236

Rossi, C., Kälin, O., Arribas, J., & Tortosa, A. (2002). Diagenesis, provenance and reservoir quality of Triassic T AGI sandstones from Ourhoud field, Berkine (Ghadames) Basin, Algeria. *Marine and Petroleum Geology, 19*, 117-142.

Saller, A., Rushton, S., Buambua, L., Inman, K., McNeil, R., & Dickson, J. A. D. (2016). Presalt stratigraphy and depositional systems in the Kwanza Basin, offshore Angola. *AAPG Bulletin, 100*(07), 1135-1164. doi:10.1306/02111615216

Schlumberger. (1972). *Log Interpretation Manual: Principles* (Vol. 1). Houston: Schlumberger Well Services Inc.

Schmidt, C., Busch, B., & Hilgers, C. (2018). *Coupling diagenesis and petrophysics in red bed siliciclastics to infer fluid flow anisotropies – Analog studies from SW Germany.* Paper presented at the EAGE Conference Proceedings, Copenhagen

Schmidt, C., Busch, B., & Hilgers, C. (2020a). Compaction and cementation control on bleaching in Triassic fluvial red beds, S-Germany. *Zeitschrift der Deutschen*

Gesellschaft für Geowissenschaften, 172(4), 523-539. doi:10.1127/zdgg/2020/0233

Schmidt, C., Busch, B., & Hilgers, C. (2020b). Lateral variations of detrital, authigenic and petrophysical properties in an outcrop analog of the fluvial Plattensandstein, Lower Triassic, S-Germany. *Journal of Applied Regional Geology, 172*(4), 541-564. doi:10.1127/zdgg/2020/0234

Scholle, P. A., & Ulmer-Scholle, D. (2003). Cements and cementation. In G. V. Middleton & et al. (Eds.), *Encyclopedia of Sediments and Sedimentary Rocks* (pp. 88-182).

Schulz-Rojahn, J. P., Seeburger, D. A., & Beacher, G. (2003). Application of glauconite morphology in geosteering and for on-site reservoir quality assessment in very finegrained sandstones: Carnarvon Basin, Australia. In R. H. Worden & S. Morad (Eds.), *Clay Mineral Cements in Sandstones* (pp. 473-488): International Association of Sedimentologists.

Selley, R. (1970). *Ancient Sedimentary Environment and their Sub-Súrface Diagnosis* (Vol. 2). US: Springer

Selley, R. C. (1998). *Elements of petroleum geology*. New York: WH freeman and company.

Short, K. C., & Stäuble, A. J. (1967). Outline of Geology of Niger Delta. *American Association of Petroleum Geologists Bulletin, 51*(5), 761-779.

Siddiqui, N. A., Rahman, A. H. A., Sum, C. W., Yusoff, W. I. W., & Ismail, M. S. b. (2017). Shallow-marine Sandstone Reservoirs, Depositional Environments, Stratigraphic Characteristics and Facies Model: A Review. *Journal of Applied Sciences, 17*(5), 212-237. doi:10.3923/jas.2017.212.237

Slanky, M. (1962). Contribution a a'etude geologique du basin sedimentaire cotier au Dahomey at du Togo. *Bureau du Recherches geologiques of Minieres, Memoir, 11*, 270.

Steiber, R. G. (1973). Optimization of shale volumes in open hole logs. *Jour. Pet. Technol., 31*, 147-162.

Tiab, D., & Donaldson, E. (2012). *Petrophysics: Theory and Practice of Measuring Reservoir Rock and Fluid Transport Properties* (Third edition ed.). United Kingdom,: Gulf Professional Publishing, Elsevier inc.

Tobin, R. C., & Schwarzer, D. (2014). Effects of sandstone provenance on reservoir quality preservation in the deep subsurface: experimental modelling of deep-water sand in the Gulf of Mexico. *Geological Society, London, Special Publications, 386*(1), 27-47. doi:10.1144/sp386.17

Tuttle, M. L. W., Charpentier, R. R., & Brownfield, M. E. (1999). *The Niger Delta Petroleum System: Niger Delta Province, Nigeria, Cameroon, and Equatorial Guinea, Africa*. Retrieved from Denver, Colorado:

Udo, O. T., & Ekweozor, C. M. (1990). Significance of Oleanane occurrence in shales of Opuama Channel Complex, Niger Delta. *Energy & Fuel, 4*(3), 248-254.

Virolle, M., Brigaud, B., Féniès, H., Bourillot, R., Portier, E., Patrier, P., Derriennic, H., & Beaufort, D. (2021). Preservation and distribution of detrital clay coats in a modern estuarine heterolithic point bar in the Gironde estuary (Bordeaux, France). *Journal of Sedimentary Research, 91*(8), 812-832. doi:10.2110/jsr.2020.146

Walderhaug, O. (2000). Modeling quartz cementation and porosity in middle Jurassic Brent Group sandstones of the Kvitebjorn Field, Northern North Sea. *AAPG*

Bulletin, 84(9), 1325-1339 doi:10.1306/A9673E96-1738-11D7-8645000102C1865D

Walker, R. G. (1992). Facies, facies models and modem stratigraphic concepts. In R. G. Walker & N. P. James (Eds.), *Facies Models: Response to Sea Level Change* (pp. 1-14). Canada: Geological Association of Canada.

Walker, T. R. (1976). Diagenetic Origin of Continental Red Beds. In H. Falke (Ed.), *The Continental Permian in Central, West, and South Europe* (pp. 240-282).

Webb, J. E. (1974). Relation of Oil Migration to Secondary Clay Cementations, Cretaceous Sandstone, Wyoming. *Bull. Amer. Assoc. Petrol. Geol., 58*, 2245-2249.

Weber, K. J. (1971). Sedimentological aspects of oil fields in the Niger Delta *Environmental Geology, Minbouw, 50*(3), 559-576.

Weber, K. J., & Daukoru, E. (1975). Petroleum Geology of the Niger Delta *Journal of Mining and Geology, 12*, 9-12.

Webster, R., & Oliver, M. A. (2007). *Geostatistics for Environmental Scientists* (Vol. 2nd Edition): John Wiley & Sons Inc., New York

Whiteman, A. (1982a). *Nigeria Its Petroleum Geology, Resources and Potential* (Vol. 1). Netherlands: Springer

Whiteman, A. (1982b). *Nigeria: Its Petroleum Geology, Resources and Potential* (A. Whiteman Ed. Vol. 1). Netherlands: Springer Netherlands.

Whiteman, A. J. (1982c). *Nigeria Its Petroleum Geology, Resources and Potential.* In Nton, M.E. (2015), Petrography, compositional characteristics and stable isotope geochemistry of the Ewekoro formation from Ibese Corehole, eastern Dahomey basin, southwestern Nigeria. Global Journal of Geological Sciences 13(1):35-52

Wimmers, K., & Koehrer, B. (2014). Integration of Sedimentology, Petrophysics and Rock Typing as Key to Understanding a Tight Gas Reservoir. *Oil Gas European Magazine, 4(0179-3187/14/IV),* OG 196-OG 200.

Wright, V. P. (1992). Paleosol recognition: A guide to early diagenesis in terrestrial settings. In K. H. Wolf & G. V. Chilingarian (Eds.), *Diagenesis III, Developments in Sedimentology 47* (pp. 591-619): Elsevier Science Publishers B.V.

Wright, V. P., & Baceta, J. I. (2020). The mantle, CO_2 and the giant Aptian chemogenic lacustrine carbonate factory of the South Atlantic: Some carbonates are made, not born. *Sedimentology, 69*(1), 47-73. doi:10.1111/sed.12835

Wyllie, M. R. J., Gregory, A. R., & Gardner, L. W. (1956). Elastic wave velocities in heterogeneous and porous media. *Geophysics, 21*(1), 41-70.

8 Appendix

8.1 Supplementary material 2.1

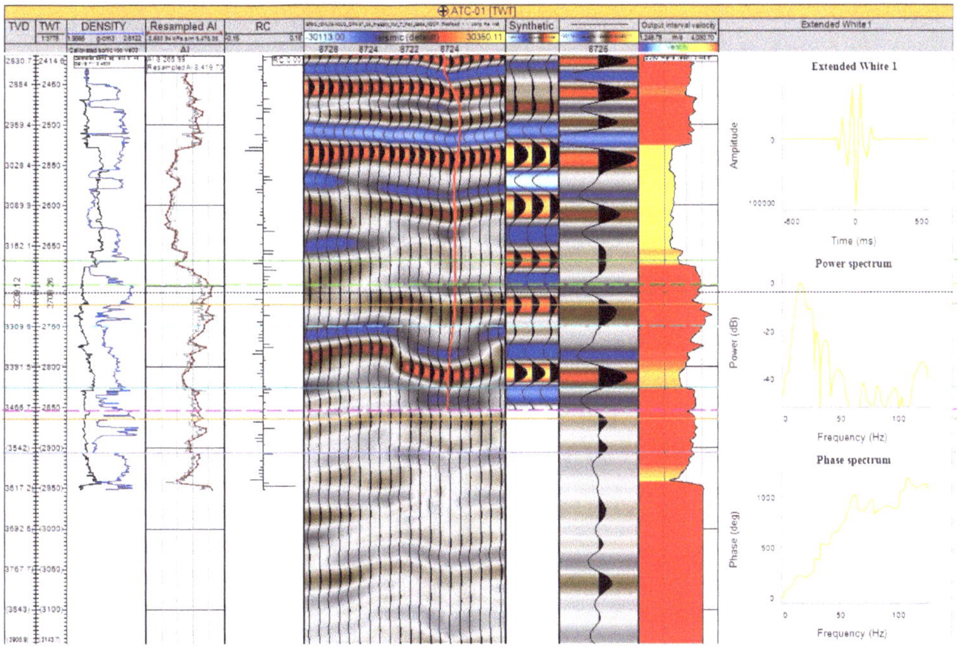

Figure 8.1. Well to seismic tie using check shot data calibrated sonic, density, and extracted wavelets from ATC-01. A time shift of 18 ms was necessary to obtain a fairly good match for the well. Track 1 = true vertical depth (TVD; m), track 2 = two-way travel time (TWT; ms) track 3 = overlay of Density and calibrated sonic logs, track 4 = Acoustic impedance (AI), track 5 = Reflection coefficient (RC), track 6 = default seismic trace, track 7 = synthetic seismogram, tract 8 = 1D seismic, track 9 = output interval velocity, track 10 = Extended White wavelet.

8.2 Supplementary material 2.2

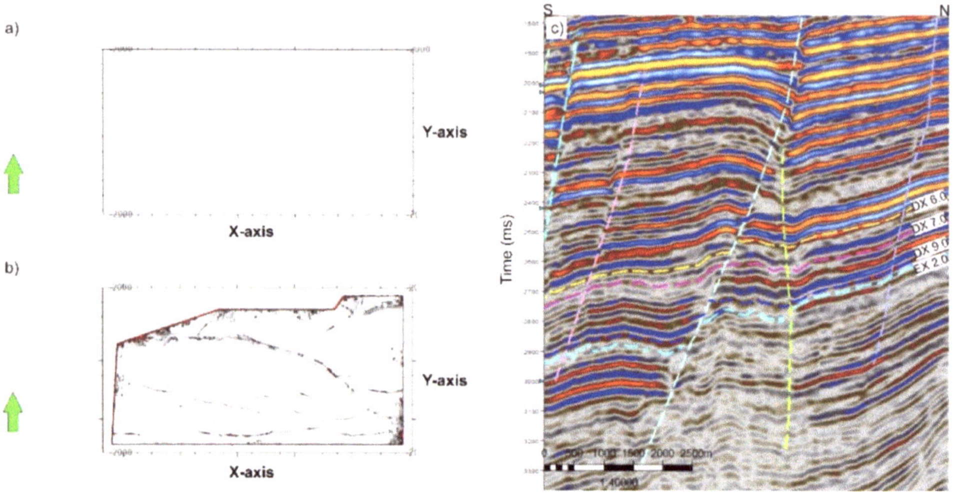

Figure 8.2. Time slice at 2032 ms showing a) ant tracking and a) variance attributes used for fault visualization and mapping. c) interpreted seismic line (In-Line 8705) showing mapped faults and horizons.

8.3 Supplementary material 2.3

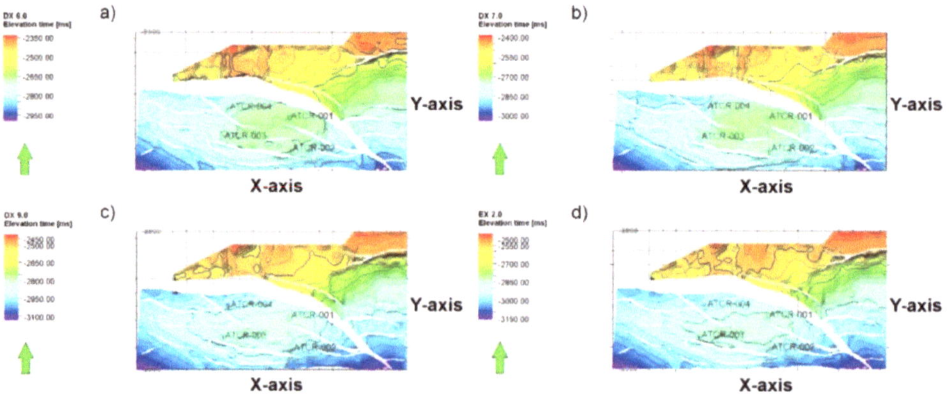

Figure 8.3. a – d) Time structure maps at the top of the identified reservoir units (DX 6.0, DX 7.0, DX 9.0, and EX 2.0).

8.4 Supplementary material 2.4

Table 8.1: Formula used for in the determination of the petrophysical parameters.

Properties	Formula	Reference	Remarks
Total porosity (ϕ_{tot})	$\frac{\rho_{mat} - \rho_{log}}{\rho_{mat} - \rho_{flu}}$	(Schlumberger, 1972)	ρ_{mat} = matrix density = 2.65 ρ_{log} = log density ρ_{flu} = fluid density = 0.9
	$\left(\frac{\Delta t - \Delta t_{mat}}{\Delta t_{flu} - \Delta t_{mat}}\right) \times \frac{1}{C_p}$	(Wyllie et al., 1956)	Δt = sonic transit time Δt_{mat} = matrix sonic Δt_{flu} = fluid sonic $C_p = \Delta t_{sha}/100$ sha = shale
Effective porosity (ϕ_{eff})	$(1 - V_{sh}) \times (\phi_{tot})$	(Asquith & Krygowski, 2004)	V_{sh} =volume of shale ϕ_{tot} =total porosity
Volume of shale (V_{sh})	$0.083 = (2^{3.71IGR} - 1)$	(Atlas, 1979)	IGR = gamma ray index
	$\text{IGR} = \left(\frac{GR_{log} - GR_{min}}{GR_{max} - GR_{min}}\right)$	(Steiber, 1973)	GR_{log} = GR log GR_{min} = minimum GR reading GR_{max} = maximum GR reading
	$\frac{RESD_{log} - RESD_{san}}{RESD_{sha} - RESD_{san}}$	(Poupon et al., 1970)	$RESD_{log}$ = resistivity log (deep) $RESD_{san}$ = resistivity in sand $RESD_{sha}$ = resistivity in shale
Formation factor (F)	a/ϕ^m	(Archie, 1942)	a = tortuosity factor or local correlation factor = 0.81 ϕ = porosity m = cementation factor
Irreducible water saturation (S_{wirr})	$(F/2000)^{1/2}$	(Archie, 1942)	F = formation factor
Water saturation (S_w)	$[F \times (R_w/R_t)]^{1/2}$	(Archie, 1942)	F = formation factor R_w = formation water resistivity R_t = resistivity of the formation
J-function water saturation (Swj)	$S_{w_j} = Pc\,(S_w) / \sigma\cos\theta \sqrt{\frac{k}{\phi}}$	(Leverett, 1941)	$S_{wj} = J - function\ water\ saturation, capilliary\ pressure, S_W = water\ saturation, \sigma\ surface\ tension, \theta\ angle\ of\ internal\ reflection, k = permeability, \phi = porosity,$
Permeability (K)	$k = 10{G_d}^2 \times \phi^{3.64+m} \times {S_i}^{-3.64}$	(Onengiyeofori et al., 2019)	$k = permeability, \phi = porosity, Gd = dominant\ grain\ size, m = cementation\ exponent, Si = sorting\ index\ of\ the\ formation\ (varying\ from\ 0.7\ to\ 1)$

8.5 Supplementary material 2.5

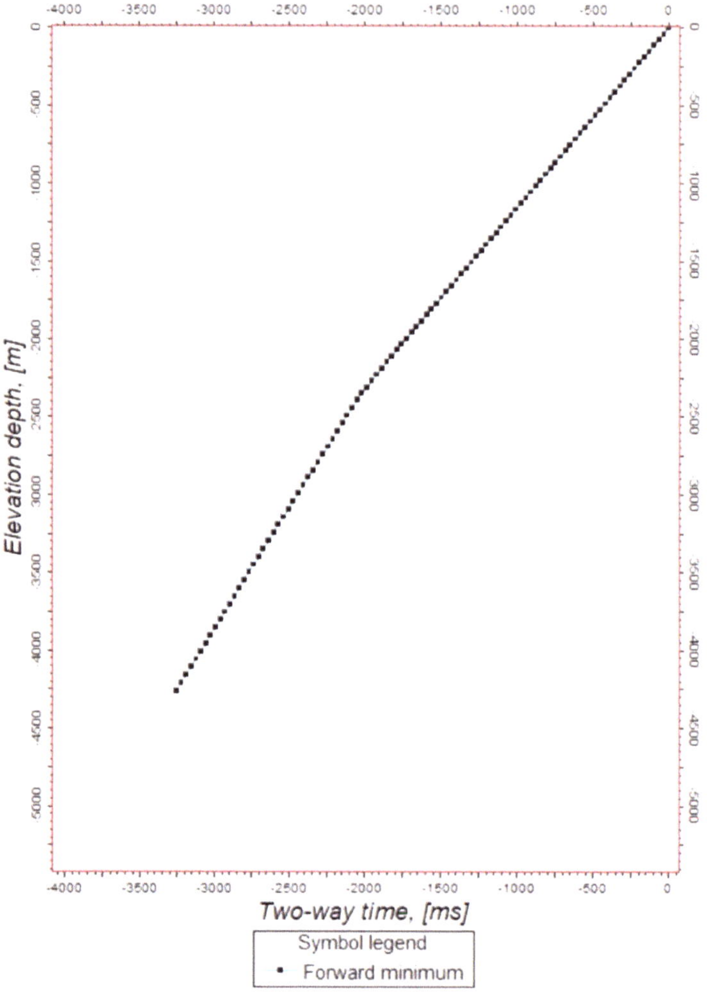

Figure 8.4. Time (TWT; ms) versus depth (m) relationship used for the depth conversion.

8.6 Supplementary material 3.1

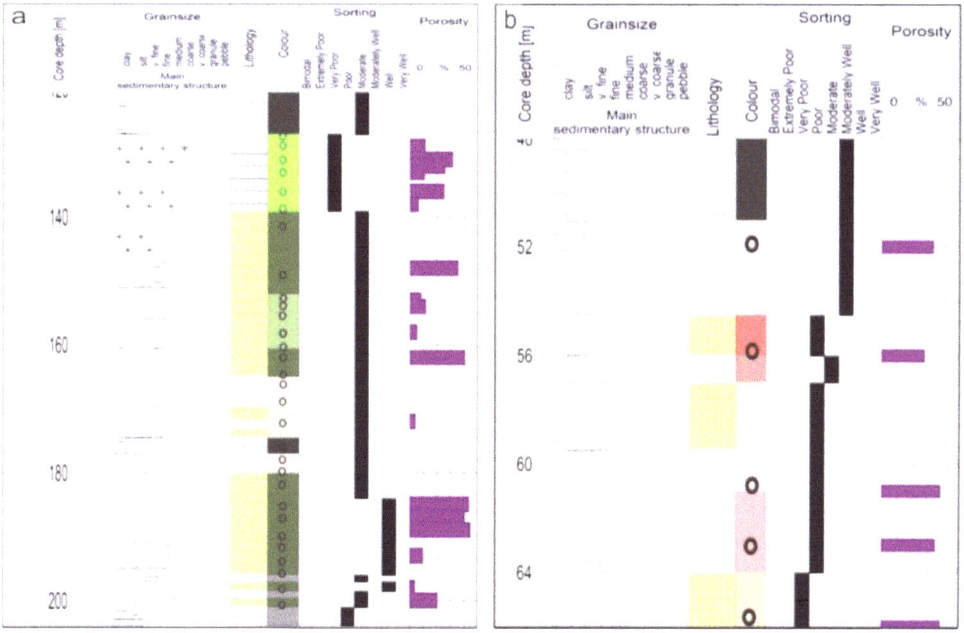

Figure 8.5 a, b. Lithological column of Ilaro (IL) and Ijagun (IJ) Wells.

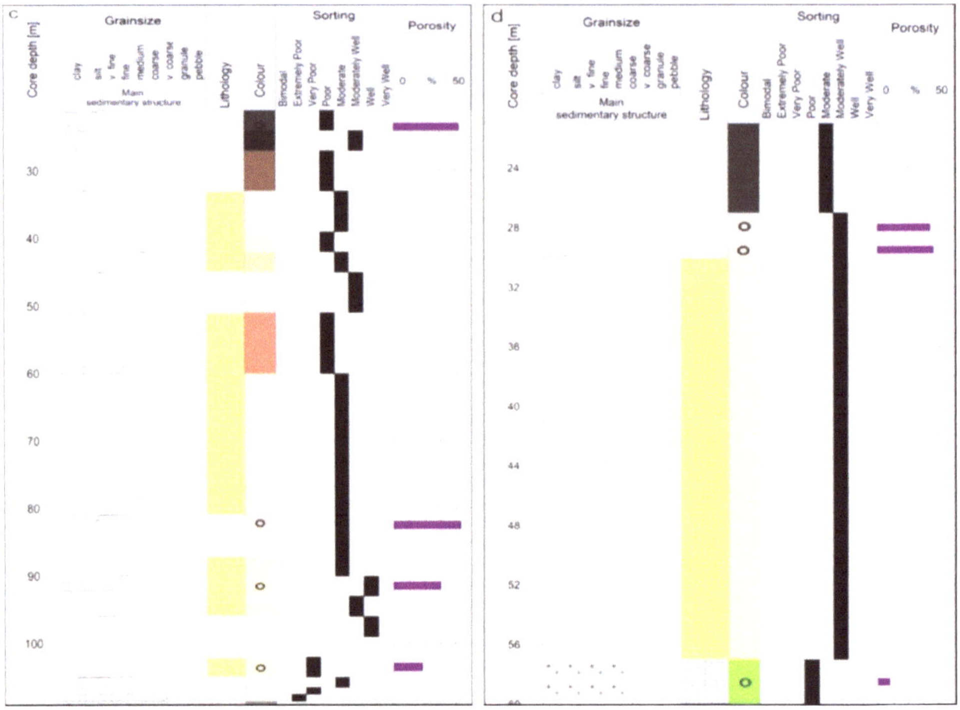

Figure 8.5 c, d. Lithological column of Akinmodiro (AM) and Akinboni-Idogun (AK) Wells

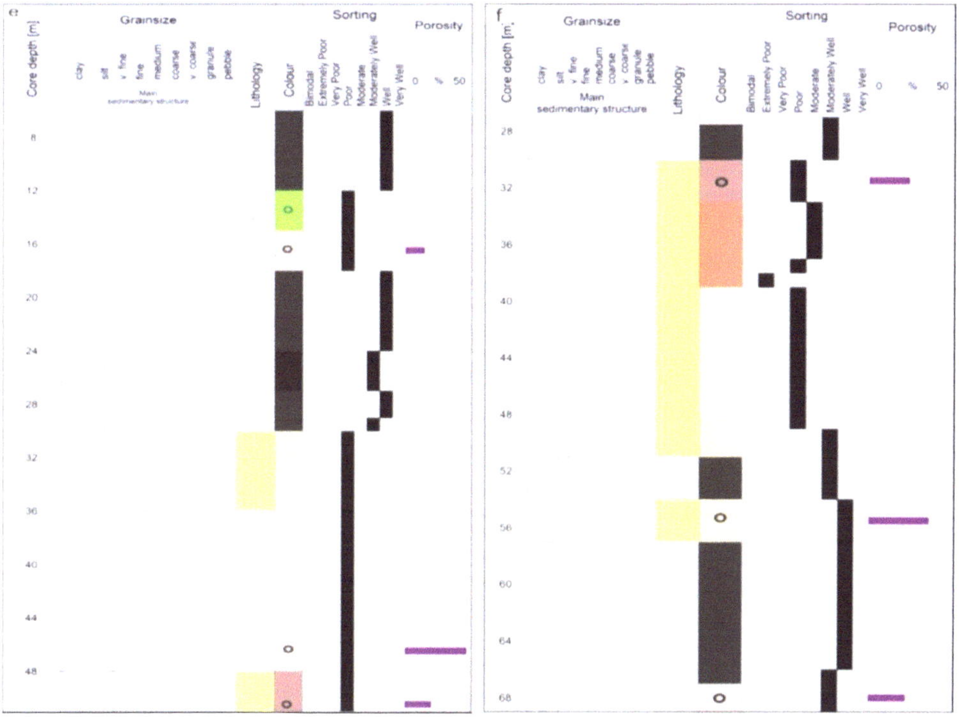

Figure 8.5 e, f. Lithological column of Ohosu I (OH) and Ohosu II (SU) Wells.

8.7 Supplementary material 3.2

Table 8.2. Point count results and petrophysical data for the studied samples. Qtz = quartz, Frag. = fragment, Argl. = argillaceous, pf. = pore filling, pl. = pore lining, cmt. = cement, IGV = intergranular volume, EW = Ewekoro, AR =Araromi, AF = Afowo.

S/N	Sample ID	Depth	Formation	Qtz. mono.	Qtz. poly.	Qtz. total	K-feldspar	Plagioclase feldspar	Total feldspar	Rock fragments	Mica	Zircon	Rutile	Tourmaline	Staurolite	Opaque	Total HM	Shell feldspar	Glauconite	Pyrite	Phosphate
		(m)		(%)	(%)	(%)	(%)	(%)	(%)	(%)	(%)	(%)	(%)	(%)	(%)	(%)	(%)	(%)	(%)	(%)	(%)
1	IL 18	136	EW	27.3	1.7	29	0	0	0	4.33	0	0	0	0	0	0	0	1	2.7	1.3	0
2	IL 21	148	EW	27.3	1.3	28.7	0	0	0	1.3	0	0	0	0	0	0	0	4.7	2.7	1	0
3	IL 22	152	EW	32.3	0.7	33	0.7	3.3	4	1	0	0	0	0	0	0	0	1.7	1	1.3	0
4	IL 24	154	EW	41.7	1.7	43.3	0	0.3	0.3	0.3	0	0	0	0	0	0	0	3	1.3	0	0
5	IL 25	158	EW	35.7	1.7	37.3	3.7	3.7	7.3	0.7	0	0	0	0	0	0	0	8	3.3	0.3	0
6	IL 27	162	AR	33.7	0.7	34.3	0.7	2.3	3	1	0	0	0	0	0	0	0	2.3	0.7	0.3	0
7	IL 31	172	AR	41	0.3	41.3	2.3	3.3	5.7	0.3	0	0	0	0.3	0	0	0.3	0.7	0.3	0	0
8	IL 32	174	AR	38.3	4	42.3	1.3	2	3.3	0.3	0	0	0	0	0	0	0	0.3	5	0	0
9	IL 35	178	AR	25	1.7	26.7	0.3	0.3	0.7	1.3	0	0	0	0	0	0	0	0.7	5.3	1.7	0.7
10	IL 36	179	AR	29.3	2.7	32	0	0.7	0.7	0.3	0	0	0	0.3	0	0	0.3	0	14	1	0
11	IL 38	184	AR	41.7	7	48.7	1.3	4	5.3	0	0.3	0	0	0	0	0	0	0	2	0	0
12	IL 39	185	AR	26.7	8	34.7	0	0.7	0.7	0.7	0	0.3	0	0	0	0	0.3	0	19.7	4.3	5
13	IL 40	187	AR	37.3	1	38.3	0	1	1	0.7	0	0.3	0	0	0	0	0.3	0	20.7	0.7	1.7
14	IL 41	189	AR	41.3	4.7	46	0.3	0.7	1	0	0	0	0	0	0	0	0	0.3	1	0.7	5.7
15	IL 43	193	AR	42.3	10.3	52.7	0.3	0.7	1	1.3	0	0	0	0	0	0	0	0	0.3	0	0.7
16	IL 44	196	AR	47	2.3	49.3	0	0.3	0.3	1	0	0	0	0	0	0	0	0	1	0.3	1
17	IL 45	197	AR	44.3	4.33	48.7	0	0.3	0.3	0.7	0	0	0	0	0.3	0	0.3	0	1.7	1	1.3
18	IL 46	198	AR	49	9.3	58.3	0	0.7	0.7	0	0	0	0	0	0	0	0	0	0.3	0	0
19	IL 47	200	AR	44.7	6.3	51	0	1	1	0	0	0	0	0	0	0	0	0	1.3	0.3	1.3
20	IL 48	201	AR	36	7	43	0	0.3	0.3	0.7	0	0	0	0	0	0	0	1.3	0.7	2	0
21	AK 01	28	AR	44.7	3	47.7	0.3	0	0.3	0.3	0	0	0	0.3	0	0	0.3	0.3	0	0.3	1.3
22	AK 02	29.5	AR	28	3	31	0	0.3	0.3	1.3	0.7	0	0	0.3	0	0	0.3	0	0	0.3	0.3
23	AM 06	52	AF	31.7	1.7	33.3	0.7	1	1.7	0.3	2.7	0	0	0	0	0.7	0.7	0	0	1.3	1.3
24	AM 08	56	AF	27.7	0.3	28	0	0.7	0.7	0.3	3.7	0	0	0	0	0	0	0.3	0	1	0
25	AM 09	61	AF	14.3	0.7	15	0	0.3	0.3	0.3	2.3	0	0	0	0	0	0	0	0	0.3	0
26	AM 10	63	AF	18.7	1	19.7	0	1	1	0.7	2	0	0	0	0	0	0	0	0	0.3	0.3
27	AM 11	66	AF	40.7	2	42.7	0	0	0	0.7	0.7	0.3	0	0	0	0.3	0.7	0	0	0	0
28	IJ 02	23.5	AF	5.7	0	5.7	0	0	0	0	1.3	0	0	0	0	0	0	0	0	0.7	0
29	IJ 06	91.5	AF	29.7	4.3	34	0	0	0	0.3	0	0.7	0	0	0	0	0.7	0.3	0	0	0
30	IJ 07	103.5	AF	36.3	16.3	52.7	0	0	0	0.3	0.3	0.3	0	0	0	0	0.3	0	0	0	0
31	IJ 08	107.5	AF	32	3.7	35.7	1.7	3	4.7	0.3	0.3	0	0	0	0	0	0	0	0	1	1
32	OH 03	16.5	AF	22	0.7	22.7	0	0	0	0	0	0	0	0	0	0	0	0	0	3.7	0.3
33	OH 04	46.5	AF	47.3	2.3	49.7	0	0	0	0.7	0	0	0	0	0	0	0	0	0	1	0.3
34	OH 05	50.5	AF	47.7	4	51.7	0.3	0	0.3	0.3	0	0	0	0	0	0	0	0	0	0.3	0.3
35	SU 02	31.5	AF	38.3	4.3	42.7	0	0	0	0.7	0	0	0	0	0	0	0	0	0	0.3	25.7
36	SU 04	55.5	AF	58.3	5	63.3	0	0	0	2.7	0	0	0	0	0	0	0	0	0	0	0
37	SU 06	68	AF	47.3	3	50.3	0	2.3	2.3	1	0	0	0	0	0	0	0	0	0	0	0

Table 8.2. (continued).

Sample ID	Kaolinitic matrix	Detrital clay rim	Smectitic clay pf.	Illite-Smectite pl.	Total matrix + Detrital Argl.	Illitic pf.	Illitic pl.	Smectite pf.	Kaolinite cement	Kaolinite rep. feldspar	Illite rep. mica	Illite rep. feldspar	Total clay cement.	Phosphatic clay cement	Limonitic clay cement	Carbonate mud	Fe. Calcite cement	Ferroan dolomitic cement	Non Ferroan Dol. cement	Siderite cement.	Carbonate cement (total)
	(%)	(%)	(%)	(%)	(%)	(%)	(%)	(%)	(%)	(%)	(%)	(%)	(%)	(%)	(%)	(%)	(%)	(%)	(%)	(%)	(%)
IL 18	0.7	0	0	0	0.7	0	0	0	0	0	0	0	0	0	0	28.3	6	0	0	0	6
IL 21	1.3	0	0	0	1.3	6.7	0	0.7	0	0	0	0	7.3	0	0	21	1	0	0	0	1
IL 22	0	0	24	0	24	0	0	0	2	0	0	0	2	0	0	0.3	0	0	0	0	0
IL 24	0.3	0	0	0	0.3	0	0	0	0	0	0	0	0	0	0	12.7	1.3	31.7	0	0	33
IL 25	0	0	0	0	0	0	0	0	0.7	0	0	0	0.7	0	0	1	2	36.7	0	0	38.7
IL 27	0	0	11.7	0	11.7	8	0.3	0	0	0	0	0	8.3	0	0	0	0	0	0	0	0
IL 31	0	0	0	0	0	0	0	0	0	0	0	0	0	0	0	42.7	6.7	1	0	0	7.7
IL 32	0	0	15.3	6	21.3	0	0	0	0	0	0	0	0	0	0	0	0	0	0	0	0
IL 35	1	0	52.7	0	53.7	0	0	0	0	0	0	0	0	0	0	0	0	0	0	0	0
IL 36	2.3	0	31	0.7	34	0	0	0	0	0	0	0	0	0	0	0	0	0	0	0	0
IL 38	0	0	15.7	1	16.7	0	0	0	0	0	0	0	0	0	0	0	0	0	0	0	0
IL 39	2.7	0	1	0.3	4	0	0	0	0	0	0	0	0	0	0	0	0	0	0	0	0
IL 40	0	0	0.7	0.3	1	0	0	0	0	0	0	0	0	0	0	0	0	0	0	0	0
IL 41	0.3	0	10.7	1	12	0	0	0	0	0	0	0	0	0	0	0	0	0	0	0	0
IL 43	0	0	0.3	0	0.3	0	0	0	0	0	0	0	0	0	0	0	37.7	0	0	0	37.7
IL 44	6	0	7.3	1.7	15	0	0	0	0	0	0	0	0	0	0	0	1	0	0	0	1
IL 45	2.3	0	15	2.3	19.7	0.3	0	0	0	0	0	0	0.3	0	0	0	0	0	0	0	0
IL 46	0	0.3	0.3	0	0.7	0	0	0	0	0	0	0	0	0	0	0	0	35.7	0.7	0	36.3
IL 47	0	0	8.7	1.3	10	0	0	0	0	0	0	0	0	0	0	0	0	30	0	0	30
IL 48	0	0	32	1.7	33.7	0	0	0	0	0	0	0	0	0	0	0	0	0	0	0	0
AK 01	0.3	0	0	0	1.7	6.3	2.3	0	0	0	0	0	8.7	0	14.3	0	0	0	0	0	0
AK 02	1	0	0	0	1	0	0	0	0	0	0	0	0	0	26	0	0	0	0	0	0
AM 06	1	0	0	0	1	0	0	0	0	0	0.7	0	0.7	0	38.3	0	0	0	0	0	0
AM 08	55.7	0	0	0	55.7	0	0	0	0	0	0	0.3	0.3	0	0.7	0	0	0	0	0	0
AM 09	71.3	0	0	0	71.3	0	0	0	0	0	0	0	0	0	0	0	0	0	0	0	0
AM 10	70	0	0	0	70	0	0	0	0	0	0.3	0	0.3	0	0	0	0	0	0	0	0
AM 11	49	0	0	0	49	0	0	0	0	0	0	0	0	0	0	0	0	0	0	0	0
IJ 02	85.3	0	0	0	85.3	0	0	0	0	0	0	0	0	4.3	0	0	0	0	0	0	0
IJ 06	61.3	0	0	0	61.3	0	0	0	0	0	0	0	0	0	0.3	0	0	0	0	0	0
IJ 07	45	0	0	0	45	0	0	0	0	0	0	0	0	0	0	0	0	0	0	0	0
IJ 08	27	0	2	0	29	19.3	0.7	0	0	1.3	1	0.3	22.7	0	0	0	0	0	0	0	0
OH 03	8	0	0	0	8	0	0	0	0	0	0	0	0	0	0	0	0	0	0	62.3	62.3
OH 04	19.7	0	0	0	19.7	0	0	0	0	0	0	0	0	0	0	0	0	0	0	0	0
OH 05	4.7	0	0	0	4.7	0	0	0	0	0	0	0	0	0	0	0	0	0	0	33.7	33.7
SU 02	3	0	0	0	3	0	0	0	0	0	0	0	0	0	16.7	0	0.7	0	0	0	0.7
SU 04	1.3	0	0	0	1.3	0	0	0	0	0	0	0	0	0	0	0	0	0	0	0	0
SU 06	1.7	0.1	0	0	1.7	0	0	0	0	0	0	0	0	0	0	0	0	0	0	26.3	26.3

Table 8.2. (continued).

Sample ID	Fracture cement	FeOx cement	Organic matter	Intergranular pore	Secondary pore	Total Optical pore	Total	Petrophysical data	Plug He- porosity	Plug air-permeability	Grain density	Additional sample data	Average grain size	Grain size class	Sorting	IGV	Microporosity
	(%)	(%)	(%)	(%)	(%)	(%)	(%)		(%)	(mD)	(g/cm3)		(mm)				(%)
IL 18	0	0	0	24	2.7	26.7	100		25.39	5.562	2.795		0.185	F	P	30	0
IL 21	0	0	0	30.3	0.7	31	100		35.85	109.23	2.691		0.145	F	P	31.3	4.85
IL 22	0	0	0	26.7	5	31.7	100		NA	NA	NA		0.188	F	P	26.7	NA
IL 24	0	0	0	5.7	0	5.7	100		12.04	1.328	2.766		0.201	F	P	38.7	6.37
IL 25	0	0	0	1.3	1.3	2.7	100		5.44	0.006	2.724		0.284	M	P	40	2.78
IL 27	0	0	0	36	2.3	38.3	100		40.76	1035.9	2.772		0.165	F	P	36	2.43
IL 31	0	0	0	0	1	1	100		4.56	0.0004	2.706		0.213	F	P	7.7	3.56
IL 32	0	0	0	27	0.3	27.3	100		NA	NA	NA		0.191	F	P	27	NA
IL 35	0	0	0	0	9.3	9.3	100		NA	NA	NA		0.11	VF	P	0	NA
IL 36	0	0	0	16	1.7	17.7	100		NA	NA	NA		0.09	VF	P	16	NA
IL 38	0	0	0	26.7	0.3	27	100		NA	NA	NA		0.231	M	P	26.7	NA
IL 39	0	0	0	30.7	0	30.7	100		43.78	248.11	2.909		0.072	VF	P	30.7	13.12
IL 40	0	0	0	35.7	0	35.7	100		40.58	218.06	2.819		0.075	VF	P	35.7	4.92
IL 41	0	0	0	33.3	0	33.3	100		45.15	377.86	2.847		0.14	F	P	33.3	11.82
IL 43	0.3	0	0	2.3	3.3	5.7	100		10.02	0.302	2.697		0.772	C	P	40	4.35
IL 44	0	0	0	27.3	2.7	30	100		NA	NA	NA		0.207	F	P	28.3	NA
IL 45	0	0	0	25	1	26	100		NA	NA	NA		0.172	F	P	25	NA
IL 46	0	0	0	0.3	3.3	3.7	100		4.26	0.009	2.698		0.332	M	P	36.7	0.59
IL 47	0	0	0	4.3	0.7	5	100		20.45	15.597	2.819		0.187	F	P	34.3	15.45
IL 48	0	0	0	16	2.3	18.3	100		NA	NA	NA		0.157	F	P	16	NA
AK 01	0	0	0.7	16.7	7.3	24	100		34.46	1764.4	2.496		0.121	F	P	31	10.46
AK 02	0	0	23.7	4	11	15	100		36.6	361.43	2.356		0.124	F	P	30	21.60
AM 06	0	0	7	10.3	1.3	11.7	100		33.4	287.28	2.449		0.08	VF	P	48.7	21.73
AM 08	0	0	0	7.7	1.7	9.3	100		27.64	0.069	2.715		0.111	F	P	8.3	18.31
AM 09	0	0	0	0.3	10	10.3	100		38.12	0.114	2.886		0.13	F	P	0.3	27.79
AM 10	0	0	0	0	5.7	5.7	100		34.85	0.111	2.784		0.123	F	P	0	29.18
AM 11	0	0	0	0.3	6	6.3	100		38	5.328	2.734		0.197	F	P	0.3	31.67
IJ 02	0	0	0	0	2.7	2.7	100		42.33	0.655	2.892		NA	NA	NA	4.3	39.67
IJ 06	0	0	0	0	3	3	100		30.31	0.059	2.703		0.243	VF	P	0.3	27.31
IJ 07	0	0	0	0	1.3	1.3	100		18.34	0.047	2.683		0.349	M	P	0	17
IJ 08	0	0	0	1	4.3	5.3	100		NA	NA	NA		NA	F	P	1	NA
OH 03	0	0	0	0.7	2.3	3	100		14.62	0.082	3.139		0.076	VF	P	63	11.62
OH 04	0	0	0	23.3	5.3	28.7	100		46.62	10000	2.822		0.111	VF	P	23.3	17.95
OH 05	0	0	0	8	0.7	8.7	100		19.86	25.65	2.998		0.214	M	P	41.7	11.19
SU 02	0	5	0	0.7	4.7	5.3	100		23.76	0.048	3.152		0.324	M	VP	23	18.42
SU 04	0	0	0	32.3	0.3	32.7	100		35.49	>10000	2.684		0.198	F	P	32.3	2.82
SU 06	0	1	0	14.3	3	17.3	100		21.86	2.033	3.163		0.117	VF	P	41.7	4.53

8.8 Supplementary material 3.3

Table 8.3. Point count result and petrophysical data of the limestones. EW = Ewekoro, AR = Araromi, AF = Afowo.

S/N	Sample ID	Depth	Formation	Foraminifera	Gastropod	Brachiopod	Other Shell undifferentiated	Echinoid	Ostracod	Pelecypod	Ferroan calcite cmt.	Ferroan dolomitic cement	Quartz mono.	Quartz poly.	Non Fe. calcite cement	Glauconite cement	Glauconite grain	Worm tube	Dolomite non Fe	Pyrite	Rock fragment
		(m)		(%)	(%)	(%)	(%)	(%)	(%)	(%)	(%)	(%)	(%)	(%)	(%)	(%)	(%)	(%)	(%)	(%)	(%)
1	IL 13	128.0	EW	0.7	0.0	0.0	1.7	0.0	0.0	0.0	0.0	74.0	1.3	0.0	0.0	0.0	20.7	0.0	0.0	0.3	1.3
2	IL 14	129.0	EW	0.0	0.3	7.3	11.0	0.0	0.0	0.0	0.0	17.7	0.3	0.0	0.0	0.0	1.3	0.3	0.0	0.0	0.0
3	IL 15	131.0	EW	5.0	3.0	4.0	9.0	0.0	0.0	1.0	4.0	56.7	0.3	0.3	0.0	3.3	2.3	0.0	0.0	0.7	0.0
4	IL 16	132.0	EW	12.0	1.0	2.0	7.7	0.3	0.0	2.7	5.3	34.0	0.7	0.3	0.0	0.7	0.3	0.0	0.0	0.7	0.0
5	IL 17	133.0	EW	4.7	3.0	2.7	9.7	0.0	1.0	1.3	0.0	50.3	0.3	0.3	0.0	0.3	0.7	0.0	0.0	1.0	0.0
6	IL 19	138.0	EW	0.7	0.7	1.3	3.7	0.0	0.0	0.0	12.0	7.7	26.3	2.0	12.0	0.7	0.3	0.0	0.0	0.3	0.0
7	IL 23	153.0	EW	0.7	0.0	0.7	2.7	0.0	0.0	0.3	0.0	1.3	26.7	1.7	0.0	0.0	0.3	0.0	0.0	0.0	0.0
8	AK 03	58.5	AR	1.0	1.0	4.0	10.0	0.0	0.3	1.7	3.0	4.7	8.3	0.0	0.0	0.0	0.3	0.0	56.0	1.3	0.0
9	OH 02	13.5	AR	0.0	1.0	1.0	4.3	0.0	0.3	0.3	2.3	0.0	4.3	0.0	0.0	0.0	0.0	0.0	57.0	2.0	1.0

Sample ID	Phosphate grain	Lime mud/ micrite	Clay mineral	Mouldic pore	Opaque undifferentiated	Intergranular por.	Sec. Por. undifferentiated	Sec Por. Cmt.	Sec Por. Shell	Total optical pore	Total	Petrophysical data	Plug He-Porosity	Plug Air-Permeability	Grain density	Microporosity
	(%)	(%)	(%)	(%)	(%)	(%)	(%)	(%)	(%)	(%)	(%)		(%)	(mD)	(g/cm3)	(%)
IL 13	0.0	0.0	0.0	0.0	0.0	0.0	0.0	0.0	0.0	0.0	100		NA	NA	NA	NA
IL 14	0.0	59.3	0.0	0.0	0.0	1.7	0.3	0.0	0.3	2.3	100		11.6	0.1	2.8	9.3
IL 15	0.0	0.0	0.0	0.0	0.0	0.0	0.0	8.3	2.0	10.3	100		31.7	12.9	2.8	21.4
IL 16	0.0	24.7	0.0	0.0	0.0	0.0	0.0	7.3	0.3	7.7	100		26.1	2.5	2.8	18.4
IL 17	0.0	23.0	0.0	0.7	0.0	0.0	0.0	0.7	0.3	1.0	100		11.0	0.0	2.8	10.0
IL 19	0.0	32.0	0.0	0.0	0.3	0.0	0.0	0.0	0.0	0.0	100		6.7	0.1	2.7	6.7
IL 23	0.0	65.7	0.0	0.0	0.0	0.0	0.0	0.0	0.0	0.0	100		8.5	0.0	2.8	8.5
AK 03	0.3	6.3	1.7	0.0	0.0	0.0	0.0	0.0	0.0	0.0	100		8.0	0.0	2.7	8.0
OH 02	1.0	15.3	9.0	0.0	0.0	0.0	0.3	0.7	0.0	1.0	100		NA	NA	NA	NA

8.9 Supplementary material 4.1

Table 8.4. Outcrop sample locations and lithological descriptions.

Location number (Sample number)	Formation	Lithology	Locality	Latitude (N)	Longitude (E)	Elevation (m)	Descriptions
L1 (AFo_01A, B)	Afowo	Oil sand	Egbe (Along Okitipupa - Ode-Aye Road)	06° 38' 14.3"	004° 48' 14.3"	42 ± 4	Fine-grained, dark grey to black, oil sand exposed along a river channel.
L2 (-)	Basement	Schist	Ladan-Aro (Along River Oluwa)	06° 39' 13.0"	004° 42' 32.6"	67 ± 5	Basement complex rock (schist) exposed along a river channel.
L3 (Afo_03A, B)	Afowo	Oil sand and shale	Idi-Obilayo (Along River Lisoro)	06° 38' 36.4"	004° 34' 47.6"	50 ± 5	Black, fine- to medium-grained tar-bearing sandstone which becomes shaly towards the top.
L4 (Afo_04)	Afowo	Oil sand and shale	Orita-J4 (Osho area, along Onikitimbi Road)	06° 40' 45.6"	004° 18' 36.5"	47 ± 4	Fine- to coarse-grained, black, tar sand. Light to dark grey shale (sometimes carbonaceous) and intercalation of claystone.
L5 (AF_05A - C, E, G, I, J, K)	Afowo	Sandstone	Ijebu-Imusin (Along Ore - Ijebu-Ode Expressway)	05° 46' 57.8"	004° 00' 08.0"	53 ± 3	Fine-grained, whitish to yellowish, generally finning upward, friable/unconsolidated with parallel beddings.
L6 (AF_06A - C)	Afowo	Sandstone	Ijagun/Mabolufon Area (Along Benin - Sagamu Expressway)	06° 47' 50.3"	003° 54' 13.7"	33 ± 3	Fine- to coarse-grained, yellowish to brownish, friable/weakly consolidated sandstone with ferruginous layers. Sometimes shows parallel beddings.
L7 (AF_07A, C - G)	Afowo	Sandstone	Ijebu-Ife (Along Ore - Ijebu-Ode Expressway)	06° 46' 20.3"	004° 01' 54.9"	44 ± 3	Fine- to medium-grained, friable, parallel-bedded sandstone. Above the sandstone is a unit of light brownish, well indurated claystone.
L8 (Afo_08)	Afowo	Oil sand	Looda (Along Irele Road)	06° 39' 07.1"	004° 53' 22.4"	43 ± 3	Fine- to medium-grained, black, tar-bearing sandstone.
L9 (AR_09A, C)	Araromi	Limestone and shale	Ayadi	06° 38' 23.0"	004° 53' 30.1"	25 ± 3	Light grey shale with interbedded, grey to dark grey, fossiliferous (with visible shells) limestone.
L10 (AFo_10B)	Afowo	Oil sand	Orisunbare_Ilubirin (Along Agbabu Road)	06° 38' 34.6"	004° 49' 44.0"	32± 4	Silty to Fine-grained, oil-bearing sandstone (exposed along a minor river stream in the swamp).
L11 (IS_11A, C - I)	Ise	Sandstone	Along Ago-Iwoye - Ilishan Road (NW of Ijesa-Ijebu)	06° 55' 01.9"	003° 47' 22.1"	47 ± 3	Medium- to coarse-grained, cross-bedded, conglomeratic/pebbly sandstone, with interbedded/intercalated Kaolinitic claystone. A weathered basement complex rock is also exposed at this location. The Ise formation unconformably overlies the basement complex rock at this location. Part of the kaolinite resulted from the in-situ weathering of feldspathic basement rock.
L12 (IS_12A, B)	Ise	Sandstone	Abeokuta (Along Abeokuta – Sagamu Expressway; adjacent PHCN Substation)	07° 06' 19.8"	003° 23' 32.6"	124 ± 4	Medium- to coarse-grained, poorly sorted, micaceous sandstone.

Table 8.4. (continued).

Location number (Sample number)	Formation	Lithology	Locality	Latitude (N)	Longitude (E)	Elevation (m)	Descriptions
L13 (IS_13)	Ise	Sandstone	Abeokuta (Along Lagos – Abeokuta Expressway; Km 10, South of ARO Psychiatric Hospital)	07° 04' 58.9"	003° 17' 35.7"	62 ± 3	Medium- to coarse-grained sandstone.
L14 (EW_14A, B)	Ewekoro	Limestone	Ibese	07° 00' 00.9"	003° 02' 19.9"	48 ± 3	Fossiliferous limestone, with interbedded shale. Exposed at Ibese quarry section (limited access granted).
O15 (IL_O15A - C)	Ilaro	Sandstone	Ilaro (Along Ilaro - Papalanto Road)	06° 53' 35.2"	003° 02' 04.5"	81 ± 3	Fine- to coarse-grained, yellowish to brownish sandstone with clay clasts which are most times dissolved.
A15 (IL_A15D - H)	Akinbo-Oshosun and Ilaro	Sandstone and Shale	Ilaro (Along Ilaro - Papalanto Road)	06° 53' 34.7"	003° 02' 03.8"	90 ± 3	Fine- to coarse-grained, yellowish to brownish sandstone with interbedded claystone (Ilaro; 0 to 15 m) and light to dark grey fissile shale (Akinbo-Oshosun; 15 to 30.5 m) recovered from about 30.5 m deep local well (hand-dug). This is located uphill (south) of the O15 outcrop.
B15 (IL_B15A - C)	Ilaro	Sandstone and Shale	Ilaro (Along Ilaro - Papalanto Road)	06° 53' 32.7"	003° 02' 03.7"	93 ± 3	Fine- to coarse-grained, yellowish cross-bedded sandstone which is sometimes ferruginous, yellowish to brownish recovered from an about 15 m deep local well (hand-dug). This is located uphill (south) of the O15 outcrop.
L16 (IL_16A - E)	Ilaro	Sandstone and claystone	Ajegunle (Along Ilaro - Papalanto Road)	06° 53' 13.3"	003° 07' 48.6"	115 ± 3	Medium- to coarse-grained, yellowish to brownish, cross-bedded sandstone with intercalated claystone and clay clasts. This is exposed in an abandoned sand quarry.
L17 (-)	Akinbo-Oshosun	Shale	Along Ijebu-Ode - Epe Expressway	06° 45' 06.9"	003° 58' 22.0"	28 ± 3	Light grey, partly fissile shale. Only exposed in drainage currently under construction.
L18 (AK_18 A, C, D)	Akinbo-Oshosun	Sandstone and claystone	Odolewu-Ijebu (Along Ijebu-Ode - Ilaro Expressway)	06° 44' 45.7"	003° 58' 42.0"	61 ± 3	Light grey to brownish, very poorly sorted (matrix-rich) sandstone and claystone.
L19 (AK_19)	Akinbo-Oshosun	Sandstone and claystone	Augustine University Area	06° 39' 16.1"	004° 01' 18.7"	26 ± 3	Light grey to whitish matrix-supported sandstone and claystone.
L20 (CBF_20A - C)	Cretaceous Benin Flank (CBF)	Sandstone	Ori-Orhin (Along Owo - Ifon Road)	06° 58' 00.5"	005° 44' 33.1"	165 ± 8	Medium- to coarse-grained, conglomeratic/pebbly, cross-bedded sandstone. Well indurated at some point with large pebbles embedded in the sand. Granitic basement complex rock is exposed at about 800 to 1 km south of this location. However, actual contact/unconformity was not found.
L21 (CBF_21)	Cretaceous Benin Flank (CBF)	Sandstone and shale	Elerin-Nla	06° 49' 10.7"	005° 40' 04.3"	87 ± 3	Medium- to coarse-grained, pebbly and ferruginous sandstone. Not well-exposed shale lies below.

Table 8.4 (continued).

Location number (Sample number)	Formation	Lithology	Locality	Latitude (N)	Longitude (E)	Elevation (m)	Descriptions
L22 (CBF_22Ai, Aii, B, C)	Cretaceous Benin Flank (CBF)	Sandstone	Okeluse Area (Along a river channel)	06° 49' 02.9"	005° 38' 35.1"	60 ± 3	Very fine- to medium-grained, ferruginous sandstone.
L23 (CBF_23A, B)	Cretaceous Benin Flank (CBF)	Sandstone	Okeluse (Okeluse Garage)	06° 47' 05.3"	005° 35' 33.7"	69 ± 3	Conglomeratic sandstone at the top, silty to claystone at the base, with a ferruginous layer (siltstone).
L24 (CBF_ 24A, B)	Cretaceous Benin Flank (CBF)	Limestone and sandstone	Okeluse (Along Omialayo River channel)	06° 46' 35.0"	005° 34' 56.2"	60 ± 3	Fine-grained (sometimes pebbly), parallel-bedded sandstone and fossiliferous, greyish limestone. No complete sequence at a single point. The sandstone is interpreted to be at the top of the limestone base of eroded sand and pebbles found in the river.
GB Well (EW_GB01 - 04)	Ewekoro	Limestone, claystone and shale	Gbokoto	06° 59' 0.4"	002° 44' 0.2"	-	At the top, the well consists of fissile greyish shale and claystone (up to 16.5 m). Between 16.5 and 35.4 m are interbedded limestone, claystone and shale. The limestone is light to dark grey and contains few visible fossils and vugs.
IG Well (EW_IG01, 02)	Ewekoro	Limestone, claystone and shale	Igbeme	6° 55' 0.8"	002° 47' 0.3"	-	The top 21 m is made up of yellowish to brownish and greyish claystone. Interbedded light grey limestones are found between 21 and 34.2 m. The limestone is light grey with visible vugs and a thin calcite vein.

8.10 Supplementary material 4.2

Table 8.5. Modal composition and sedimentological parameters of the studied sandstone samples. Os = oil sand, Sst = sandstone, Akin.-Osh. = Akinbo-Oshosun, CBF = Cretaceous Benin Flank, _C = coarse, _M = medium, _F = fine.

S/N	Sample ID	Formation	Lithology	Detrital phases (%)	Total quartz grain	Plagioclase grain	K- feldspar grain	Total feldspar	Metamorphic RF, undifferentiated	Igneous RF	Siltstone RF	Claystone/shale RF	Lithoclasts, undifferentiated.	Mica	Heavy mineral	FeOx grain	Detrital matrix	Illuviated/infiltrated clay
1	AFo_01A	Afowo	Os		56.33	0.00	0.00	0.00	0.00	0.00	0.00	0.00	0.00	0.00	0.00	0.33	14.67	0.00
2	AFo_01B	Afowo	Os		56.67	0.00	0.00	0.00	0.00	0.00	0.00	0.00	0.00	0.33	0.00	0.33	15.67	0.00
3	AFo_03A	Afowo	Os		55.00	0.00	0.00	0.00	0.00	0.00	0.00	0.00	0.67	1.67	0.00	1.00	8.00	0.00
4	AFo_3B	Afowo	Os		49.00	0.00	0.00	0.00	0.00	0.00	0.00	0.00	0.33	1.00	0.00	1.00	12.67	0.00
5	AFo_04	Afowo	Os		65.33	0.00	0.00	0.00	0.00	0.00	1.00	0.00	0.00	0.33	0.00	0.00	5.33	0.00
6	AF_05A	Afowo	Sst		59.67	0.00	0.00	0.00	0.00	0.00	0.33	0.33	0.00	0.00	0.67	0.00	8.67	0.00
7	AF_05B	Afowo	Sst		65.33	0.00	0.00	0.00	0.00	0.00	0.00	0.33	0.00	0.00	1.67	0.33	7.67	0.00
8	AF_05C	Afowo	Sst		55.67	0.00	0.00	0.00	0.00	0.00	0.00	0.00	0.00	0.00	0.33	0.00	1.33	0.00
9	AF_05E	Afowo	Sst		61.67	0.00	0.00	0.00	0.00	0.00	0.33	0.33	0.00	0.33	0.67	0.00	1.33	0.00
10	AF_05G	Afowo	Sst		64.33	0.00	0.33	0.33	0.00	0.00	0.00	0.00	0.00	0.00	0.00	0.67	0.67	0.00
11	AF_05I	Afowo	Sst		55.67	0.00	0.00	0.00	0.00	0.00	0.00	0.33	0.00	0.67	0.67	0.00	3.00	0.00
12	AF_05J	Afowo	Sst		61.00	0.00	0.00	0.00	0.00	0.00	1.00	0.33	0.33	0.67	0.33	0.00	6.67	0.00
13	AF_05K	Afowo	Sst		63.00	0.00	0.00	0.00	0.00	0.00	0.33	0.00	0.33	0.00	1.00	1.00	31.33	0.00
14	AF_06A	Afowo	Sst		69.33	0.00	0.00	0.00	0.00	0.00	0.00	0.00	0.00	0.00	0.00	0.00	1.33	0.00
15	AF_06B	Afowo	Sst		71.33	0.00	0.00	0.00	0.00	0.00	0.00	0.00	0.00	0.00	0.00	0.00	1.67	0.00
16	AF_06C	Afowo	Sst		60.00	0.00	0.00	0.00	0.00	0.00	0.33	0.00	0.00	0.00	0.00	0.33	13.67	0.00
17	AF_07A	Afowo	Sst		58.00	0.00	0.00	0.00	0.00	0.00	0.00	0.33	0.33	2.00	2.00	0.67	1.33	0.00
18	AF_07C	Afowo	Sst		58.67	0.00	0.00	0.00	0.00	0.00	0.00	1.00	0.00	2.67	1.00	0.00	0.67	0.00
19	AF_07D	Afowo	Sst		53.67	0.00	0.00	0.00	0.00	0.00	0.00	0.33	0.33	1.67	0.33	0.00	23.33	0.00
20	AF_07E	Afowo	Sst		62.00	0.00	0.00	0.00	0.00	0.00	0.00	0.33	0.00	1.33	0.33	0.33	0.67	0.00
21	AF_07F	Afowo	Sst		48.00	0.00	0.00	0.00	0.00	0.00	0.00	0.00	0.00	1.00	0.33	0.33	22.33	0.00
22	AF_07G	Afowo	Sst		42.67	0.00	0.00	0.00	0.00	0.00	0.00	1.00	0.00	0.67	0.00	0.33	34.00	0.00
23	AFo_08	Afowo	Os		67.33	0.00	0.00	0.00	0.00	0.00	0.33	0.00	0.00	0.00	0.00	0.00	6.33	0.00
24	AFo_10B	Afowo	Os		57.67	0.00	0.00	0.00	0.00	0.00	0.33	0.00	0.00	0.00	0.00	0.33	7.00	0.00
25	IS_11A	Ise	Sst		48.67	0.33	0.00	0.33	0.00	0.00	0.00	0.00	0.00	0.00	0.00	0.67	18.33	0.00
26	IS_11C	Ise	Sst		59.67	0.00	0.00	0.00	0.00	0.00	0.00	0.00	0.00	0.00	1.00	3.00	13.67	2.67
27	IS_11D	Ise	Sst		65.67	0.00	0.00	0.00	0.00	0.00	0.00	0.00	0.00	0.33	0.00	0.00	7.00	0.00
28	IS_11E	Ise	Sst		64.00	0.00	0.00	0.00	0.00	0.00	0.00	0.33	0.00	1.33	0.00	0.33	9.00	0.67
29	IS_11F	Ise	Sst		51.67	0.00	0.00	0.00	0.00	0.00	0.67	2.33	0.33	4.00	0.00	0.33	13.33	4.00
30	IS_11G	Ise	Sst		53.33	0.00	0.00	0.00	0.00	0.00	0.00	2.00	0.00	0.67	0.33	0.33	23.67	10.33
31	IS_11H	Ise	Sst		58.67	0.00	0.00	0.00	0.00	0.00	0.00	0.00	0.00	1.00	0.00	1.00	5.00	0.00
32	IS_11I	Ise	Sst		37.33	0.00	0.00	0.00	0.00	0.00	0.00	0.33	0.00	0.00	0.33	1.33	38.00	0.33
33	IS_12A	Ise	Sst		50.00	0.33	0.00	0.33	0.00	0.00	0.00	0.00	0.00	0.00	0.33	0.00	2.00	29.67
34	IS_12B	Ise	Sst		54.67	0.00	0.00	0.00	0.00	0.00	0.33	0.00	0.00	2.67	0.00	0.00	3.33	18.33
35	IS_13	Ise	Sst		69.33	0.00	0.00	0.00	0.00	0.00	0.00	0.00	0.00	0.00	0.00	0.33	13.33	0.00
36	IL_O15A	Ilaro	Sst		62.33	0.00	0.00	0.00	0.00	0.33	0.00	0.00	0.00	0.00	0.00	1.00	3.33	1.33
37	IL_O15B	Ilaro	Sst		55.67	0.00	0.00	0.00	0.00	0.00	0.33	0.00	0.00	0.00	0.00	0.00	0.00	14.67
38	IL_O15C	Ilaro	Sst		65.33	0.00	0.00	0.00	0.00	0.00	0.33	0.00	0.00	0.00	0.00	0.33	0.67	8.67
39	IL_A15D	Ilaro	Sst		40.33	0.33	0.67	1.00	0.00	0.00	0.67	0.33	0.33	0.00	0.00	0.33	25.00	0.00
40	IL_A15E	Ilaro	Sst		64.33	0.00	0.33	0.33	0.00	0.00	0.00	0.00	0.00	0.00	0.00	0.00	16.67	0.00
41	IL_A15F	Ilaro	Sst		59.00	0.00	0.00	0.00	0.00	0.00	0.00	0.00	0.00	0.00	0.33	0.00	16.67	0.00
42	IL_A15G	Ilaro	Sst		51.00	0.00	0.00	0.00	0.00	0.00	0.00	0.67	0.00	0.00	0.00	0.00	9.33	0.00
43	IL_A15H	Ilaro	Sst		59.33	0.33	0.00	0.33	0.00	0.00	0.67	0.00	0.00	0.00	0.00	0.33	4.67	5.67
44	IL_B15A	Ilaro	Sst		61.00	0.67	0.00	0.67	0.00	0.00	0.33	0.00	0.00	0.00	0.00	0.00	4.00	0.00
45	IL_B15B	Ilaro	Sst		59.00	0.00	0.00	0.00	0.00	0.00	0.00	0.33	0.33	0.00	0.00	0.33	8.67	0.00
46	IL_B15C	Ilaro	Sst		56.67	0.00	0.00	0.00	0.00	0.00	0.00	0.00	0.00	0.00	0.00	0.33	0.00	11.00
47	IL_16A	Ilaro	Sst		49.00	0.00	0.00	0.00	0.00	0.00	0.00	0.67	0.33	0.00	0.00	0.00	9.67	18.00
48	IL_16B	Ilaro	Sst		54.67	0.00	0.00	0.00	0.00	0.00	0.00	0.00	0.00	0.00	0.33	0.00	5.67	19.33
49	IL_16C	Ilaro	Sst		52.67	0.00	0.00	0.00	0.00	0.00	0.33	1.00	0.00	0.33	0.00	0.33	9.33	17.00
50	IL_16D	Ilaro	Sst		58.33	0.00	0.33	0.33	0.00	0.00	0.33	0.00	0.00	0.00	0.33	0.00	8.00	0.33
51	IL_16E	Ilaro	Sst		63.00	0.67	0.00	0.67	0.00	0.00	0.67	0.00	0.00	0.00	0.00	0.00	6.67	0.00

Table 8.5. (continued).

S/N	Sample ID	Formation	Lithology	Detrital phases (%)	Total quartz grain	Plagioclase grain	K- Feldspar grain	Total feldspar	Metamorphic RF, undifferentiated	Igneous RF	Siltstone RF	Claystone/shale RF	Lithoclasts, undifferentiated.	Mica	Heavy mineral	FeOx grain	Detrital matrix	Illuviated/infiltrated clay
52	AK_18A	Akin.-Osh.	Sst		46.33	0.00	0.00	0.00	0.00	0.00	0.00	0.00	0.00	0.00	0.00	0.33	47.00	0.33
53	AK_18C	Akin.-Osh.	Sst		45.33	0.00	0.00	0.00	0.00	0.00	0.00	2.33	0.00	0.00	0.00	1.00	10.00	6.67
54	AK_18D	Akin.-Osh.	Sst		53.33	0.00	0.00	0.00	0.00	0.00	0.00	0.00	0.33	0.00	0.00	0.33	1.33	1.00
55	AK__19	Akin.-Osh.	Sst		41.67	0.00	0.00	0.00	0.00	0.00	0.00	2.33	0.00	0.00	0.33	0.33	43.33	0.00
56	CBF-C_20A	CBF_C	Sst		64.67	0.00	0.00	0.00	0.00	0.00	0.00	0.00	0.00	0.00	0.00	0.00	11.67	5.33
57	CBF-C_20B	CBF_C	Sst		72.00	0.00	0.00	0.00	0.00	0.00	0.00	0.00	0.33	0.00	0.00	0.00	9.00	1.00
58	CBF-20C	CBF_C	Sst		62.67	0.00	0.00	0.00	0.00	0.00	0.33	0.00	0.00	0.00	0.67	0.00	5.67	1.33
59	CBF-C_21	CBF_C	Sst		43.33	1.33	2.33	3.67	0.00	0.00	0.00	0.00	0.00	0.00	0.00	0.33	0.33	0.00
60	CBF-M_22Ai	CBF_M	Sst		69.33	0.00	0.00	0.00	0.00	0.00	0.00	0.00	0.33	0.00	0.00	0.00	0.33	0.00
61	CBF-M_22Aii	CBF_M	Sst		75.67	0.00	0.33	0.33	0.00	0.00	0.33	0.00	0.00	0.33	0.00	0.00	5.67	0.00
62	CBF-M_22B	CBF_M	Sst		61.33	0.00	0.00	0.00	0.00	0.00	0.33	0.00	0.00	0.00	0.00	0.00	2.00	0.00
63	CBF-M_22C	CBF_M	Sst		53.00	0.00	0.00	0.00	0.00	0.00	0.00	0.00	0.00	0.00	0.00	3.33	0.00	0.00
64	CBF-F_23A	CBF_F	Sst		60.00	0.00	0.00	0.00	0.00	0.00	0.33	0.00	0.00	0.00	0.00	0.67	1.67	0.33
65	CBF-F_23B	CBF_F	Sst		58.00	0.00	0.00	0.00	0.00	0.00	0.00	0.67	0.33	0.00	0.00	0.00	32.33	0.00
66	CBF-F_24B	CBF_F	Sst		63.00	0.00	0.00	0.00	0.00	0.00	0.00	0.00	0.00	0.00	0.00	0.00	3.67	0.00
Minimum, maximum and average																		
		Afowo	Os	Min.	49	0	0		0	0	0	0	0	0	0	0	5.33	0
				Max.	67.33	0	0		0	0	1	0	0.67	1.67	0	1	15.67	0
				Avg.	58.19	0	0		0	0	0.24	0	0.14	0.48	0	0.43	9.95	0
		Afowo	Sst	Min.	42.67	0	0		0	0	0	0	0	0	0	0	0.667	0
				Max.	71.33	0	0.33		0	0	1	1	0.33	2.67	2	1	34	0
				Avg.	59.41	0	0.02		0	0	0.14	0.28	0.08	0.65	0.55	0.24	9.39	0
		Ise	Sst	Min.	37.33	0	0		0	0	0	0	0	0	0	0	2	0
				Max.	69.33	0.33	0		0	0	0.67	2.33	0.33	4	1	3	38	29.67
				Avg.	55.73	0.06	0		0	0	0.09	0.46	0.03	0.91	0.18	0.67	13.33	6
		Ilaro	Sst	Min.	40.33	0	0		0	0	0	0	0	0	0	0	0	0
				Max.	65.33	0.67	0.67		0	0.33	0.67	1	0.33	0.33	0.33	1	25	19.33
				Avg.	56.98	0.125	0.083		0	0.021	0.229	0.188	0.063	0.021	0.06	0.19	8.02	6
		Akin.-Osh.	Sst	Min.	41.67	0	0		0	0	0	0	0	0	0	0.33	1.33	0
				Max.	53.33	0	0		0	0	0	2.333	0.333	0	0.33	1	47	6.67
				Avg.	46.67	0	0		0	0	0	1.167	0.083	0	0.08	0.5	25.42	2
		CBF-C	Sst	Min.	43.33	0	0		0	0	0	0	0	0	0	0	0.33	0
				Max.	72	1.333	2.333		0	0	0.333	0	0.333	0	0.67	0.33	11.67	5.33
				Avg.	60.67	0.333	0.583		0	0	0.083	0	0.083	0	0.17	0.08	6.67	1.92
		CBF-M	Sst	Min.	53	0	0		0	0	0	0	0	0	0	0	0	0
				Max.	75.67	0	0.333		0	0	0.333	0	0.333	0.333	0	3.3	5.67	0
				Avg.	64.83	0	0.083		0	0	0.167	0	0.083	0.083	0	0.83	2	0
		CBF-F	Sst	Min.	58	0	0		0	0	0	0	0	0	0	0	1.67	0
				Max.	63	0	0		0	0	0.333	0.667	0.333	0	0	0.67	32.33	0.33
				Avg.	60.33	0	0		0	0	0.111	0.222	0.111	0	0	0.22	12.56	0.11

Table 8.5. (continued).

Sample ID	Authigenic phases (%)	Glauconite	Pyrite	Goethite	Hematite	Undifferentiated FeOx cement	Kaolinite cement	Bitumen	Porosity (%)	Intergranular porosity	Intragranular/secondary porosity	Total	Calculations (%)	Pre-emplacement porosity	IGV	QFR (%)	Quartz
AFo_01A		0.00	0.33	0.00	0.00	0.00	0.00	13.00		12.67	2.67	100.00		25.67	25.67		100.00
AFo_01B		0.00	0.00	0.00	0.00	0.00	0.00	15.33		9.67	2.00	100.00		25.00	25.00		100.00
AFo_03A		0.00	0.33	0.00	0.00	0.00	0.00	17.33		11.33	4.67	100.00		28.67	28.67		98.80
AFo_3B		0.00	1.00	0.00	0.67	0.00	0.00	23.00		7.00	4.33	100.00		30.00	30.67		99.32
AFo_04		0.00	0.00	0.00	0.00	0.00	0.00	6.67		20.67	0.67	100.00		27.33	27.33		98.49
AF_05A		0.00	0.00	0.00	0.00	0.33	0.00	0.00		29.67	0.33	100.00		-	30.00		98.90
AF_05B		0.00	1.00	0.00	0.00	0.00	0.00	0.00		23.67	0.00	100.00		-	23.67		99.49
AF_05C		0.00	1.00	0.00	0.00	19.67	0.00	0.00		21.33	0.67	100.00		-	41.00		100.00
AF_05E		0.00	0.00	0.00	0.00	0.00	0.00	0.00		34.00	1.33	100.00		-	34.00		98.93
AF_05G		0.00	0.00	0.00	0.00	0.00	0.00	0.00		28.67	5.33	100.00		-	28.67		99.48
AF_05I		0.00	0.00	0.00	0.00	0.00	0.00	0.00		37.67	2.00	100.00		-	37.67		99.40
AF_05J		0.00	0.00	0.00	0.00	0.00	0.00	0.00		29.67	0.00	100.00		-	29.67		97.34
AF_05K		0.00	0.00	0.33	0.00	0.00	0.00	0.00		0.33	2.33	100.00		-	0.67		98.95
AF_06A		0.00	0.00	0.00	6.67	3.67	0.00	0.00		18.33	0.67	100.00		-	25.00		100.00
AF_06B		0.00	0.00	0.00	14.67	0.00	0.00	0.00		12.33	0.00	100.00		-	27.00		100.00
AF_06C		0.00	0.33	0.00	0.00	0.00	0.00	0.00		19.67	5.67	100.00		-	19.67		99.45
AF_07A		0.00	0.33	0.00	0.00	0.00	0.00	0.00		31.00	4.00	100.00		-	31.00		98.86
AF_07C		0.00	0.00	0.00	0.00	0.00	0.00	0.00		33.33	2.67	100.00		-	33.33		98.32
AF_07D		0.00	0.33	0.00	0.33	0.00	0.00	0.00		19.33	0.33	100.00		-	19.67		98.77
AF_07E		0.00	0.33	0.00	0.00	0.00	0.00	0.00		32.00	2.67	100.00		-	32.00		99.47
AF_07F		0.00	0.33	1.67	0.00	0.00	0.00	0.00		25.67	0.33	100.00		-	27.33		100.00
AF_07G		0.00	0.00	2.67	1.00	0.00	0.00	0.00		17.67	0.00	100.00		-	21.33		97.71
AFo_08		0.00	1.00	0.00	0.33	0.00	0.00	13.00		5.67	6.00	100.00		18.67	19.00		99.51
AFo_10B		0.00	0.33	0.00	1.00	0.00	0.00	13.00		7.33	13.00	100.00		20.33	21.33		99.43
IS_11A		0.00	0.00	3.00	0.00	0.00	0.00	0.00		29.00	0.00	100.00		-	32.00		99.32
IS_11C		0.00	0.00	0.00	6.33	0.00	1.67	0.00		11.33	0.67	100.00		-	22.00		100.00
IS_11D		0.00	0.00	0.00	0.00	0.00	1.33	0.00		25.33	0.33	100.00		-	26.67		100.00
IS_11E		0.00	0.00	1.67	0.00	0.00	0.33	0.00		21.00	1.33	100.00		-	23.67		99.48
IS_11F		0.00	0.33	0.00	0.33	0.00	0.67	0.00		21.33	0.67	100.00		-	26.33		93.94
IS_11G		0.00	0.33	0.00	0.00	0.00	0.00	0.00		7.33	1.67	100.00		-	17.67		96.39
IS_11H		0.00	0.00	4.00	11.33	0.00	1.33	0.00		17.00	0.67	100.00		-	33.67		100.00
IS_11I		0.00	0.00	0.00	16.67	0.00	0.00	0.00		2.00	3.67	100.00		-	19.00		99.12
IS_12A		0.00	0.00	1.00	0.00	0.00	0.00	0.00		16.67	0.00	100.00		-	47.33		99.34
IS_12B		0.00	0.00	0.33	0.00	0.00	0.00	0.00		20.33	0.00	100.00		-	39.00		99.39
IS_13		0.00	0.00	0.00	1.00	0.00	0.00	0.00		16.00	0.00	100.00		-	17.00		100.00
IL_O15A		0.00	0.00	0.00	10.67	0.00	0.00	0.00		21.00	0.00	100.00		-	33.00		99.47
IL_O15B		0.00	0.00	0.00	7.33	0.00	0.00	0.00		21.00	1.00	100.00		-	43.00		99.40
IL_O15C		0.00	0.00	0.00	9.33	0.00	0.00	0.00		15.33	0.00	100.00		-	33.33		99.49
IL_A15D		14.33	0.00	0.00	0.00	5.67	0.00	0.00		10.67	1.33	100.00		-	16.33		94.53
IL_A15E		0.00	0.00	0.00	0.00	0.00	0.00	0.00		18.67	0.00	100.00		-	18.67		99.48
IL_A15F		0.00	0.00	0.00	0.00	0.00	0.00	0.00		24.00	0.00	100.00		-	24.00		100.00
IL_A15G		0.00	0.33	10.00	0.33	0.00	0.00	0.00		28.33	0.00	100.00		-	38.67		98.71
IL_A15H		0.00	0.00	4.67	8.00	0.00	0.00	0.00		16.33	0.00	100.00		-	34.67		98.34
IL_B15A		0.00	0.00	0.00	7.00	0.00	0.00	0.00		27.00	0.00	100.00		-	34.00		98.39
IL_B15B		0.00	0.00	5.00	0.33	0.00	0.00	0.00		26.00	0.00	100.00		-	31.33		98.88
IL_B15C		0.00	0.00	14.00	1.67	0.33	0.00	0.00		14.67	1.33	100.00		-	41.67		100.00
IL_16A		0.00	0.00	0.00	0.00	0.00	0.00	0.00		22.33	0.00	100.00		-	40.33		98.00
IL_16B		0.00	0.00	0.33	0.00	0.00	0.00	0.00		19.00	0.67	100.00		-	38.67		100.00
IL_16C		0.00	0.00	0.33	0.00	0.00	0.00	0.00		18.33	0.33	100.00		-	35.67		97.53
IL_16D		0.00	0.00	0.00	0.00	0.33	0.00	0.00		31.67	0.33	100.00		-	32.33		98.87
IL_16E		0.00	0.00	1.33	10.00	0.00	0.00	0.00		17.00	0.67	100.00		-	28.33		97.93

Table 8.5. (continued).

Sample ID		Glauconite	Pyrite	Goethite	Hematite	Undifferentiated FeOx cement	Kaolinite cement	Bitumen		Intergranular porosity	Intragranular/secondary porosity	Total		Pre-emplacement porosity	IGV		Quartz
AK_18A	Authigenic phases (%)	0.00	0.00	0.00	0.00	3.33	0.00	0.00	Porosity (%)	0.33	2.33	100.00	Calculations (%)	-	4.00	QFR (%)	100.00
AK_18C		0.00	0.00	0.00	24.00	0.00	0.00	0.00		6.33	4.33	100.00		-	37.00		95.10
AK_18D		0.00	0.33	5.33	16.33	0.33	0.00	0.00		19.00	2.33	100.00		-	42.00		99.38
AK_19		0.00	0.33	0.00	4.33	0.00	0.00	0.00		4.67	2.67	100.00		-	9.00		94.70
CBF-C_20A		0.00	0.00	0.33	0.00	0.00	0.00	0.00		17.67	0.33	100.00		-	23.33		100.00
CBF-C_20B		0.00	0.00	0.00	0.33	0.00	0.00	0.00		15.67	1.67	100.00		-	17.00		99.54
CBF-20C		0.00	0.00	10.00	4.00	0.33	0.00	0.00		14.00	1.00	100.00		-	29.67		99.47
CBF-C_21		0.00	0.00	22.00	4.67	1.00	0.00	0.00		19.33	5.33	100.00		-	47.00		92.20
CBF-M_22Ai		0.00	0.00	16.67	3.00	2.00	0.00	0.00		7.00	1.33	100.00		-	28.67		99.52
CBF-M_22Aii		0.00	0.00	0.00	4.00	0.00	0.00	0.00		13.67	0.00	100.00		-	17.67		99.13
CBF-M_22B		0.00	0.00	25.33	7.67	0.00	0.00	0.00		2.67	0.67	100.00		-	35.67		99.46
CBF-M_22C		0.00	0.00	18.67	11.33	4.67	0.00	0.00		9.00	0.00	100.00		-	43.67		100.00
CBF-F_23A		0.00	0.00	1.33	16.00	0.00	0.00	0.00		19.00	0.67	100.00		-	36.67		99.45
CBF-F_23B		0.00	0.00	0.00	0.00	0.00	0.00	0.00		7.67	1.00	100.00		-	7.67		98.31
CBF-F_24B		0.00	0.00	7.33	6.00	0.00	0.00	0.00		16.67	3.33	100.00		-	30.00		100.00
Minimum, maximum and averaged																	
Afowo	Min.	0	0	0	0	0	0	6.66667		5.67	0.67			18.67	19		98.49
	Max.	0	1	0	1	0	0	23		20.67	13			30	30.67		100
	Avg.	0	0.43	0	0.29	0	0	14.47		10.62	4.76			25.10	25.38		99.36
Afowo	Min.	0	0	0	0	0	0	0		0.33	0			-	0.67		97.34
	Max.	0	1	2.67	14.67	19.67	0	0		37.67	5.667			-	41		100
	Avg.	0	0.22	0.27	1.33	1.39	0	0		24.37	1.67			-	27.16		99.12
Ise	Min.	0	0	0	0	0	0	0		2	0			-	17		93.9394
	Max.	0	0.33	4	16.67	0	1.67	0		29	3.67			-	47.33		100
	Avg.	0	0.06	0.91	3.24	0	0.48	0		17.03	0.82			-	27.67		98.82
Ilaro	Min.	0	0	0	0	0	0	0		10.67	0			-	16.33		94.53
	Max.	14.33	0.33	14	10.67	5.67	0	0		31.67	1.33			-	43		100
	Avg.	0.90	0.02	2.23	3.42	0.40	0	0		20.70	0.35			-	32.75		98.69
Akin.-Osh.	Min.	0	0	0	0	0	0	0		0.33	2.33			-	4		94.70
	Max.	0	0.33	5.33	24	3.33	0	0		19	4.33			-	42		100
	Avg.	0	0.17	1.33	11.17	0.92	0	0		7.58	2.92			-	23		97.30
CBF-C	Min.	0	0	0	0	0	0	0		14	0.33			-	17		92.20
	Max.	0	0	22	4.67	1	0	0		19.33	5.33			-	47		100
	Avg.	0	0	8.08	2.25	0.33	0	0		16.67	2.08			-	29.25		97.80
CBF-M	Min.	0	0	0	3	0	0	0		2.67	0			-	17.67		99.13
	Max.	0	0	25.33	11.33	4.67	0	0		13.67	1.33			-	43.67		100
	Avg.	0	0	15.17	6.5	1.67	0	0		8.08	0.5			-	31.42		99.53
CBF-F	Min.	0	0	0	0	0	0	0		7.67	0.67			-	7.67		98.31
	Max.	0	0	7.33	16	0	0	0		19	3.33			-	36.67		100
	Avg.	0	0	2.89	7.33	0	0	0		14.44	1.67			-	24.78		99.25

Table 8.5. (continued). n. = near, s. c. = strongly coarse, c. = coarse, f. = fine.

Sample ID	Feldspar	Rock fragments (%)		Minimum grain size (mm)	Maximum grain size (mm)	Grain size standard deviation	Average grain size (mm)	Mean grain size (mm)	Grain size class	Sorting value	Sorting class	Skewness value	Skewed class
AFo_01A	0.00	0.00	Sedimentology	0.053	0.642	0.070	0.169	0.159	fine sand	1.27	poorly sorted	-0.0474	n.-symmetrical
AFo_01B	0.00	0.00		0.064	0.485	0.073	0.175	0.159	fine sand	1.28	poorly sorted	0.05516	n.-symmetrical
AFo_03A	0.00	1.20		0.027	0.758	0.137	0.209	0.162	fine sand	1.57	poorly sorted	0.02799	n.-symmetrical
AFo_3B	0.00	0.68		0.036	0.308	0.035	0.093	0.088	very fine sand	1.28	poorly sorted	-0.0554	n.-symmetrical
AFo_04	0.00	1.51		0.019	2.367	0.421	0.721	0.697	coarse sand	1.60	poorly sorted	-0.3462	s. c.-skewed
AF_05A	0.00	1.10		0.021	0.481	0.053	0.111	0.101	very fine sand	1.31	poorly sorted	0.004	n.-symmetrical
AF_05B	0.00	0.51		0.009	1.020	0.151	0.149	0.104	fine sand	2.18	very poorly sorted	-0.1392	c.- skewed
AF_05C	0.00	0.00		0.012	0.280	0.055	0.078	0.06	very fine sand	1.51	poorly sorted	0.14289	f.-skewed
AF_05E	0.00	1.07		0.036	0.394	0.060	0.140	0.13	fine sand	1.32	poorly sorted	-0.0717	n.-symmetrical
AF_05G	0.52	0.00		0.053	2.277	0.375	0.591	0.54	coarse sand	1.63	poorly sorted	-0.2353	c. skewed
AF_05I	0.00	0.60		0.060	0.417	0.054	0.138	0.131	fine sand	1.22	poorly sorted	-0.0222	n.-symmetrical
AF_05J	0.00	2.66		0.038	0.377	0.047	0.124	0.120	very fine sand	1.21	poorly sorted	-0.1551	c. skewed
AF_05K	0.00	1.05		0.011	1.195	0.122	0.104	0.073	very fine sand	1.54	poorly sorted	0.11731	f.-skewed
AF_06A	0.00	0.00		0.127	0.752	0.109	0.311	0.306	medium sand	1.25	poorly sorted	-0.2015	c. skewed
AF_06B	0.00	0.00		0.053	1.356	0.322	0.510	0.426	coarse sand	1.72	poorly sorted	-0.1105	c. skewed
AF_06C	0.00	0.55		0.030	1.483	0.232	0.318	0.258	medium sand	1.84	poorly sorted	-0.0829	n.-symmetrical
AF_07A	0.00	1.14		0.031	0.981	0.122	0.207	0.193	fine sand	1.46	poorly sorted	-0.1597	c. skewed
AF_07C	0.00	1.68		0.019	1.404	0.207	0.325	0.276	medium sand	1.51	poorly sorted	-0.0507	n.-symmetrical
AF_07D	0.00	1.23		0.033	0.920	0.183	0.281	0.237	medium sand	1.65	poorly sorted	-0.0781	n.-symmetrical
AF_07E	0.00	0.53		0.055	1.513	0.243	0.414	0.390	medium sand	1.54	poorly sorted	-0.2318	c. skewed
AF_07F	0.00	0.00		0.031	0.603	0.104	0.155	0.128	fine sand	1.35	poorly sorted	0.0918	n.-symmetrical
AF_07G	0.00	2.29		0.074	0.298	0.042	0.158	0.154	fine sand	1.19	poorly sorted	-0.0814	n.-symmetrical
AFo_08	0.00	0.49		0.012	0.278	0.051	0.149	0.145	fine sand	1.22	poorly sorted	-0.0885	n.-symmetrical
AFo_10B	0.00	0.57		0.019	0.258	0.043	0.121	0.115	very fine sand	1.28	poorly sorted	-0.0441	n.-symmetrical
IS_11A	0.68	0.00		0.083	2.271	0.411	0.587	0.469	coarse sand	1.54	poorly sorted	0.04556	n.-symmetrical
IS_11C	0.00	0.00		0.073	2.361	0.436	0.597	0.478	coarse sand	1.57	poorly sorted	-0.0159	n.-symmetrical
IS_11D	0.00	0.00		0.053	0.798	0.147	0.330	0.304	medium sand	1.33	poorly sorted	-0.021	n.-symmetrical
IS_11E	0.00	0.52		0.080	1.052	0.186	0.339	0.279	medium sand	1.33	poorly sorted	0.24901	f.-skewed
IS_11F	0.00	6.06		0.087	2.352	0.297	0.347	0.271	medium sand	1.37	poorly sorted	0.16635	f.-skewed
IS_11G	0.00	3.61		0.047	2.363	0.401	0.411	0.313	medium sand	1.59	poorly sorted	-0.0624	n.-symmetrical
IS_11H	0.00	0.00		0.112	1.187	0.198	0.416	0.378	medium sand	1.39	poorly sorted	-0.0447	n.-symmetrical
IS_11I	0.00	0.88		0.034	1.573	0.234	0.283	0.246	medium sand	1.80	poorly sorted	-0.184	c. skewed
IS_12A	0.66	0.00		0.084	2.306	0.424	0.552	0.407	coarse sand	1.57	poorly sorted	0.14865	f.-skewed
IS_12B	0.00	0.61		0.091	1.148	0.185	0.352	0.321	medium sand	1.47	poorly sorted	-0.1058	c. skewed
IS_13	0.00	0.00		0.065	2.340	0.403	0.477	0.359	medium sand	1.54	poorly sorted	0.08981	n.-symmetrical
IL_O15A	0.00	0.53		0.073	1.796	0.260	0.420	0.339	medium sand	1.49	poorly sorted	0.12492	f.-skewed
IL_O15B	0.00	0.60		0.147	2.378	0.375	0.675	0.581	coarse sand	1.50	poorly sorted	-0.0181	n.-symmetrical
IL_O15C	0.00	0.51		0.118	1.368	0.228	0.494	0.452	medium sand	1.29	poorly sorted	0.00461	n.-symmetrical
IL_A15D	2.34	3.13		0.047	0.287	0.036	0.125	0.123	very fine sand	1.21	poorly sorted	-0.1737	c. skewed
IL_A15E	0.52	0.00		0.038	1.506	0.243	0.573	0.562	coarse sand	1.33	poorly sorted	-0.205	c. skewed
IL_A15F	0.00	0.00		0.064	1.331	0.280	0.486	0.423	medium sand	1.47	poorly sorted	-0.0369	n.-symmetrical
IL_A15G	0.00	1.29		0.081	0.865	0.140	0.311	0.286	medium sand	1.29	poorly sorted	-0.0529	n.-symmetrical
IL_A15H	0.55	1.10		0.143	1.741	0.198	0.396	0.365	medium sand	1.28	poorly sorted	-0.0244	n.-symmetrical
IL_B15A	1.08	0.54		0.080	1.144	0.186	0.439	0.419	medium sand	1.32	poorly sorted	-0.1078	c. skewed
IL_B15B	0.00	1.12		0.112	1.350	0.245	0.589	0.564	coarse sand	1.42	poorly sorted	-0.1412	c. skewed
IL_B15C	0.00	0.00		0.053	1.643	0.332	0.582	0.585	coarse sand	1.56	poorly sorted	-0.4376	s. c.-skewed
IL_16A	0.00	2.00		0.071	1.483	0.311	0.492	0.466	medium sand	1.59	poorly sorted	-0.3588	s. c.-skewed
IL_16B	0.00	0.00		0.090	1.798	0.260	0.407	0.362	medium sand	1.55	poorly sorted	-0.185	c. skewed
IL_16C	0.00	2.47		0.057	1.842	0.244	0.410	0.349	medium sand	1.38	poorly sorted	0.05791	n.-symmetrical
IL_16D	0.56	0.56		0.036	2.058	0.320	0.372	0.279	medium sand	1.75	poorly sorted	0.06357	n.-symmetrical
IL_16E	1.04	1.04		0.070	2.261	0.365	0.550	0.473	coarse sand	1.50	poorly sorted	-0.0319	n.-symmetrical

Table 8.5. (continued).

Sample ID	Feldspar	Rock fragments (%)	Sedimentology	Minimum grain size (mm)	Maximum grain size (mm)	Grain size standard deviation	Average grain size (mm)	Mean grain size (mm)	Grain size class	Sorting value	Sorting class	Skewness value	Skewed class
AK_18A	0.00	0.00	**Sedimentology**	0.027	0.953	0.111	0.122	0.096	very fine sand	1.53	poorly sorted	0.01147	n.-symmetrical
AK_18C	0.00	4.90		0.035	2.364	0.435	0.386	0.22	medium sand	2.13	very poorly sorted	0.1392	f.-skewed
AK_18D	0.00	0.62		0.058	2.230	0.313	0.435	0.358	medium sand	1.49	poorly sorted	-0.0293	n.-symmetrical
AK_ 19	0.00	5.30		0.074	1.612	0.144	0.225	0.202	fine sand	1.34	poorly sorted	0.00129	n.-symmetrical
CBF-C_20A	0.00	0.00		0.082	2.165	0.359	0.544	0.441	coarse sand	1.44	poorly sorted	0.07873	n.-symmetrical
CBF-C_20B	0.00	0.46		0.114	2.252	0.406	0.562	0.433	coarse sand	1.37	poorly sorted	0.23444	f.-skewed
CBF-20C	0.00	0.53		0.037	1.607	0.254	0.398	0.342	medium sand	1.50	poorly sorted	-0.01	n.-symmetrical
CBF-C_21	7.80	0.00		0.048	1.777	0.323	0.632	0.604	coarse sand	1.42	poorly sorted	-0.1869	c. skewed
CBF-M_22Ai	0.00	0.48		0.049	1.252	0.191	0.378	0.367	medium sand	1.30	poorly sorted	-0.3197	s. c.-skewed
CBF-M_22Aii	0.44	0.44		0.071	1.034	0.157	0.374	0.383	medium sand	1.26	poorly sorted	-0.368	s. c.-skewed
CBF-M_22B	0.00	0.54		0.054	0.763	0.120	0.345	0.334	medium sand	1.20	poorly sorted	-0.0738	n.-symmetrical
CBF-M_22C	0.00	0.00		0.046	0.314	0.048	0.148	0.144	fine sand	1.23	poorly sorted	-0.1251	coarse skewed
CBF-F_23A	0.00	0.55		0.074	0.298	0.042	0.158	0.154	fine sand	1.19	poorly sorted	-0.0814	n.-symmetrical
CBF-F_23B	0.00	1.69		0.024	0.945	0.144	0.200	0.163	fine sand	1.37	poorly sorted	0.12891	f.-skewed
CBF-F_24B	0.00	0.00		0.056	0.361	0.050	0.163	0.155	fine sand	1.22	poorly sorted	0.03034	n.-symmetrical
Minimum, maximum and average													
Afowo: Min.	0	0		0.01	0.26	0.03	0.09	0.09		1.22		-0.35	
Max.	0	1.51		0.06	2.37	0.42	0.72	0.70		1.60		0.06	
Avg.	0	0.63		0.03	0.73	0.12	0.23	0.22		1.36		-0.07	
Afowo: Min.	0	0		0.01	0.28	0.04	0.08	0.06		1.19		-0.23	
Max.	0.52	2.66		0.13	2.28	0.38	0.59	0.54		2.18		0.14	
Avg.	0.03	0.85		0.04	0.93	0.15	0.24	0.21		1.50		-0.07	
Ise: Min.	0	0		0.03	0.80	0.15	0.28	0.25		1.33		-0.18	
Max.	0.68	6.06		0.11	2.36	0.44	0.60	0.48		1.80		0.25	
Avg.	0.12	1.06		0.07	1.80	0.30	0.43	0.35		1.50		0.02	
Ilaro: Min.	0	0		0.04	0.29	0.04	0.12	0.12		1.21		-0.44	
Max.	2.34	3.13		0.15	2.38	0.37	0.67	0.58		1.75		0.12	
Avg.	0.38	0.93		0.08	1.55	0.25	0.46	0.41		1.43		-0.10	
Akin.-Osh: Min.	0	0		0.03	0.95	0.11	0.12	0.10		1.34		-0.03	
Max.	0	5.30		0.07	2.36	0.43	0.43	0.36		2.13		0.14	
Avg.	0	2.70		0.05	1.79	0.25	0.29	0.22		1.63		0.03	
CBF-C: Min.	0	0		0.04	1.60	0.25	0.40	0.34		1.37		-0.19	
Max.	7.80	0.53		0.11	2.25	0.41	0.63	0.60		1.50		0.23	
Avg.	1.95	0.25		0.07	1.95	0.34	0.53	0.45		1.43		0.03	
CBF-M: Min.	0	0		0.05	0.31	0.05	0.15	0.14		1.20		-0.37	
Max.	0.44	0.54		0.07	1.25	0.19	0.38	0.38		1.30		-0.07	
Avg.	0.11	0.36		0.06	0.84	0.13	0.31	0.31		1.25		-0.22	
CBF-F: Min.	0	0		0.02	0.30	0.04	0.16	0.15		1.19		-0.08	
Max.	0	1.69		0.07	0.94	0.14	0.20	0.16		1.37		0.13	
Avg.	0	0.74		0.05	0.53	0.08	0.17	0.16		1.26		0.03	

8.11 Supplementary material 4.3

Table 8.6. Composition of the studied limestone samples. CBF = Cretaceous Benin Flank.

S/N	Sample ID	Formation	Bioclast (%)	Foraminifera	Gastropod	Brachiopod	Echinoid	Ostracod	Undifferentiated bioclast	Other detrital fragments(%)	Carbonate grain	Phosphate	Detrital quartz	Clay/shale rock fragment	Lithoclasts undifferentiated	Kaolinitic matrix	Other undifferentiated clay matrix
1	AR_09A	Araromi		1.67	0	0.33	0	0	5.33		0	1	2	0	0.67	1	9.33
2	AR_09C	Araromi		0	0	1	0.33	0	3.33		0	1.67	0.67	0	0.33	0	2.33
3	EW_14A	Ewekoro		2	0	2.67	0.33	2	11.33		0	0	0	0	0	0	0.33
4	EW_14B	Ewekoro		1.33	0	1	3.67	0.33	2.33		0.33	0	6	0	1	1	11
5	CBF_24A	CBF		0	0	2	0	1.33	5.67		0	0.67	0	0	0	0	0
6	EW_GB01	Ewekoro		2	0	0	0	0	0.33		0	0	0	0	0	0	0
7	EW_GB02	Ewekoro		0	0	0	0	0	0		0	0	0	0	0	0	1.33
8	EW_GB03	Ewekoro		1	1	0.67	0.67	0	4.67		2	0.33	0	0	0.67	8.67	0.33
9	EW_GB04	Ewekoro		1.67	0.67	0.33	0	0	3		0.67	0	0	0	0	5	0
10	EW_IG01	Ewekoro		4	0.33	0	0	0	0.67		0	0	0	0	0	0	0.33
11	EW_IG02	Ewekoro		1	0	0	0	0	2		0	0	0	3	3.67	1	5
Minimum, maximum and average																	
		Araromi	Minimum	0	0	0.33	0	0	3.33		0	1	0.67	0	0.33	0	2.33
			Maximum	1.667	0	1	0.333	0	5.333		0	1.67	2	0	0.67	1	9.33
			Average	0.83	0	0.67	0.17	0	4.33		0	1.33	1.33	0	0.5	0.5	5.83
		CBF		0	0	2	0	1.33	5.67		0	0.67	0	0	0	0	0
		Ewekoro	Minimum	0	0	0	0	0	0		0	0	0	0	0	0	0
			Maximum	4	1	2.67	3.67	2	11.33		2	0.33	6	3	3.67	8.67	11
			Average	1.63	0.25	0.58	0.58	0.29	2.79		0.38	0.04	0.75	0.38	0.67	1.96	2.29

Sample ID	Authigenic phases (%)	Sparry calcite	Sparry Fe-calcite	Micritic calcite	Micritic Fe-calcite	Radiaxial fibrous calcite	Calcite rim cement	Fe-calcite rim cement	Dolomite	Fe-dolomite	Dolomite rhombs	Siderite	Ooid	Pelloid	Pyrite	Glauconite
AR_09A		19.67	30.33	4	11.67	0	1	1.33	1	0	0	0	0	0	0.33	0
AR_09C		34.33	12	30.33	7.33	0	4	0	0	0	0	0.33	0	0	0.67	0
EW_14A		47	3	23.33	0	0	0	0	0.67	0.67	0	0	0	2	0	0.33
EW_14B		1.67	0	0.67	0	0	0	0	0.33	0.33	40	0	0	0	4.33	2
CBF_24A		4.33	15.33	1.33	23.67	0	0	0	0	0	1.33	3.67	0	0	0.67	0
EW_GB01		68.3	1	0	0	0	2.33	0	20	0	0	5	0	0	1	0
EW_GB02		40.33	18	0	0	7.67	0.67	0	10.67	0	0	21	0	0	0.33	0
EW_GB03		50	0	9.67	0	0	0	0	0	0	0	11.33	0	0	1.67	1
EW_GB04		10.33	48	3	3.67	0	0.67	0	14	1.33	1.33	0	0.33	0.33	2.33	1
EW_IG01		73.33	6.33	0	0	0	2	0	1	0	0	0	0	0	0.33	0.33
EW_IG02		69.33	0	8.33	0	0	0	0	0	0	0.33	0.33	0	0	0	0.33
Minimum, maximum and average																
Minimum		19.67	12	4	7.33	0	1	0	0	0	0	0	0	0	0.33	0
Maximum		34.33	30.33	30.33	11.67	0	4	1.33	1	0	0	0.33	0	0	0.67	0
Average		27	21.17	17.17	9.5	0	2.5	0.67	0.5	0	0	0.17	0	0	0.5	0
		4.33	15.33	1.33	23.67	0	0	0	0	0	1.33	3.67	0	0	0.67	0
Minimum		0	0	0	0	0	0	0	0	0	0	0	0	0	0	0
Maximum		73.33	48	23.33	3.67	7.67	2.33	0	20	1.33	40	21	0.33	2	4.33	2
Average		38.79	9.54	4.42	0.46	0.96	0.71	0	5.83	0.29	5.21	4.71	0.04	0.29	1.25	0.63

Table 8.6. (continued).

Sample ID	Porosity (%)	Intergranular porosity	Secondary porosity	Secondary porosity; in lithoclasts undifferentiated	Secondary porosity; in cement	Secondary porosity; in shell	Total optical porosity	Total
AR_09A		0	0	0	4	2.33	6.33	100
AR_09C		0	0	0	0.33	0	0.33	100
EW_14A		0	0	0	0.67	3.33	4	100
EW_14B		0	0.67	0.33	0.67	0.33	2	100
CBF_24A		0.33	0	0	0	0	0.33	100
EW_GB01		0	0	0	0	0	0	100
EW_GB02		0	0	0	0	0	0	100
EW_GB03		0	0.33	0.33	2	3.67	6.33	100
EW_GB04		0	0	1.33	1	0	2.33	100
EW_IG01		0	0	0	0	0	0	100
EW_IG02		0	1	0	4	0.67	5.67	100
Minimum, maximum and average								
Minimum		0	0	0	0.33	0	0.33	
Maximum		0	0	0	4	2.33	6.33	
Average		0	0	0	2.17	1.17	3.33	
		0.33	0	0	0	0	0.33	
Minimum		0	0	0	0	0	0	
Maximum		0	1	1.33	4	3.67	6.33	
Average		0	0.25	0.25	1.04	1	2.54	